Strive for a 5:

Preparing for the
AP® Environmental Science Examination

Strive for a 5:

Preparing for the AP® Environmental Science Examination

to accompany

Friedland and Relyea
Environmental Science for the AP® course

Courtney Mayer
Northside ISD

Elisa McCracken
Brandeis High School

ISBN-13: 978-1-4641-5616-8
ISBN-10: 1-4641-5616-6

Printed in the United States of America

Fourth printing 2017

W. H. Freeman and Company
41 Madison Avenue
New York, NY 10010
Houndmills, Basingstoke RG21 6XS, England
www.whfreeman.com

CONTENTS

PREFACE

This workbook, *Strive for a 5: Preparing for the AP® Environmental Science Examination*, is designed for use with *Friedland and Relyea Environmental Science for the AP® course,* Second Edition. It is intended to help you assess your knowledge of the material covered in the textbook, to reinforce the key concepts, and to prepare you to take the AP® environmental science exam.

Study Guide

The study guide section is organized by chapter. As each chapter is covered in your class, you can use the study guide to help you identify and reinforce the important concepts in environmental science. Each chapter contains the following features:

- **Chapter and Module Opening Material:** Before you read the chapter, preview the chapter's key topics, the opening case study, and "Do the Math" boxes. Module exercises begin with a list of learning objectives.
- **Preview Key Terms/Define Key Terms:** Do these exercises before and as you read each module to help you learn important key terms.
- **Study the Figure:** Answer these questions to review important chapter concepts and practice the skill of interpreting visual representations of scientific data.
- **Review Key Terms:** At the end of each module, test your knowledge of new key terms.
- **Practice the Math:** Do these problems to reinforce the important math skills introduced in the "Do the Math" boxes in the chapter.
- **Check Your Understanding:** At the end of the chapter, review and practice the important concepts by answering these questions.
- **Practice for Free-Response Questions:** Answer these questions to help you remember and apply chapter material, in preparation to answer full free-response questions.
- **Review and Reflect:** Solidify your knowledge of chapter material and create exam study tools you will use throughout the course.
- **Unit Multiple-Choice Review Exams:** Test your understanding and retention at the end of each unit with a multiple-choice exam containing with 20-30 questions

Answers to exercises, problems, and exams can be found at the end of the book.

Test Preparation

After the chapter-by-chapter study section, you will find two full-length exams with 100 multiple-choice and 4 free-response questions, like the actual exam. After completing the practice exams, check the answers and look closely at the questions you answered incorrectly. Be sure to spend extra time with the material that you find challenging.

Best of luck with your course this year and on the exam!

Courtney Meyer
Elisa McCracken

Chapter (1) Environmental Science: Studying the State of Our Earth

Chapter Summary

This chapter introduces the study of environmental science. It discusses key concepts used in the study of environmental science, including the systems perspective, environmental indicators, sustainability, and scientific method. The chapter establishes important foundations for chapters to come. The chapter consists of 3 modules:

- **Module 1:** Environmental Science
- **Module 2:** Environmental Indicators and Sustainability
- **Module 3:** Scientific Method

Chapter Opening Case: *To Frack, Or Not to Frack*

The chapter opening case introduces you to the costs and benefits of using hydraulic fracturing, known as fracking, to extract oil and gas. This case demonstrates how human activities that are initially perceived as causing little harm to the environment can in fact have substantial adverse effects. It also illustrates the controversial side of issues that environmental scientists explore and the difficulty in obtaining absolute answers to environmental problems and questions.

Do the Math

This chapter contains the following "Do the Math" boxes to help prepare you for calculation questions you might encounter on the exam.

- "Converting Between Hectares and Acres" (page 11)
- "Rates of Forest Clearing" (page 14)

To make sure you understand the concepts and techniques presented in these boxes, do the practice problems presented in the text as well as the additional "Practice the Math" problems that appear in Module 2 of this study guide.

Module 1: Environmental Science

BEFORE YOU READ THE MODULE

Focus on Learning Objectives

Use the module learning objectives to guide your reading. On a separate piece of paper, write down each objective and take notes to help you meet each learning objective. After studying this module, you should be able to:

- define the field of environmental science and discuss its importance.
- identify ways in which humans have altered and continue to alter our environment.

Preview Key Terms

In a notebook or on a separate sheet of paper, create a table like the one shown here to help with learning new key terms in the module. Before you read, fill out the "Prediction" column. Write what you think the term might mean or what it makes you think about. Use examples from your everyday life. There are no wrong answers!

Key Term	Prediction	Definition
Write key term here.	Write what you think the term means in this column.	Define the term here. Add an example and use it in a sentence.

Key Terms

Fracking
Environment
Environmental science

Ecosystem
Biotic
Abiotic

Environmentalist
Environmental Studies

WHILE YOU READ THE MODULE

Define Key Terms

When you come across a new key term while reading the module, copy the definition into the "Definition" column of your key terms table. Add an example and use the term in a sentence. Compare your initial ideas to the actual definition.

Study the Figure

Examine Figure 1.2, "Systems within systems" on page 5. This figure demonstrates the complexity of environmental systems. Researchers can often define systems within systems, depending on the topic of study.

    1. After studying this figure, describe a set of systems within systems in the area where you live or somewhere you have visited.

AFTER YOU READ THE MODULE

Review Key Terms

Match the key terms on the left with the definitions on the right.

_____1. Fracking

a. Field of study that includes environmental science, environmental policy, economics, literature, and ethics

_____2. Environment

b. Living

_____3. Environmental Science

c. Nonliving

_____4. Ecosystem

d. A person participates in environmentalism, a social movement that seeks to protect the environment through lobbying, activism, and education

_____5. Biotic

e. A particular location on Earth distinguished by its mix of interacting biotic and abiotic components

_____6. Abiotic

f. The field of study that looks at interactions among human systems and those found in nature

_____7. Environmentalist

g. The sum of all the conditions surrounding us that influence life

_____8. Environmental Studies

h. A method of oil and gas extraction that uses high-pressure fluids to open cracks in rocks deep underground

Module 2: Environmental Indicators and Sustainability

BEFORE YOU READ THE MODULE

Focus on Learning Objectives

Use the module learning objectives to guide your reading. On a separate piece of paper, write down each objective and take notes to help you meet each learning objective. After studying this module, you should be able to:

- identify key environmental indicators and their trends over time.
- define sustainability and explain how it can be measured using the ecological footprint.

Preview Key Terms

In a notebook or on a separate sheet of paper, create a table like the one shown here to help with learning new key terms in the module. Before you read, fill out the "Prediction" column. Write what you think the term might mean or what it makes you think about. Use examples from your everyday life. There are no wrong answers!

Key Term	Prediction	Definition
Write key term here.	*Write what you think the term means in this column.*	*Define the term here. Add an example and use it in a sentence.*

Key Terms

Ecosystem service
Environmental indicator
Biodiversity
Genetic
Diversity
Species

Species
diversity
Speciation
Background
Extinction rate
Greenhouse gases

Anthropogenic
Development
Sustainability
Sustainable development
Biophilia
Ecological footprint

WHILE YOU READ THE MODULE

Define Key Terms

When you come across a new key term while reading the module, copy the definition into the "Definition" column of your key terms table. Add an example and use the term in a sentence. Compare your initial ideas to the actual definition.

Study the Figure

Use Figure 2.5, "Changes in average global temperature and in atmospheric CO_2 concentrations" on page 12 to answer the following questions:

1. According to the graph, what is the correlation between increased CO_2 levels and temperature over the past 100 years?

2. Determine approximate percent increase for carbon dioxide from 1800 to 1900 and from 1900 to 2000.

3. Compare your calculations from question 2. What is the correlation between atmospheric carbon and global temperatures?

4. Identify two possible human causes for the difference between the two centuries.

Practice the Math: Converting Between Hectares and Acres

Read "Do the Math: Converting Between Hectares and Acres" on page 11. Try "Your Turn." For more math practice, do the following problems. Remember to show your work. Use a separate sheet of paper if necessary.

2.5 acres = 1 hectare (ha)
1 acre = 0.40 ha

    Convert the following from acres to hectares.

    50,000 acres = _____ hectares

    75,000 acres = _____ hectares

    150,000 acres = _____ hectares

Practice the Math: Rates of Forest Clearing

Read "Do the Math: Rates of Forest Clearing" on page 14. Try "Your Turn." For more math practice, do the following exercise. Remember to show your work. Use a separate sheet of paper if necessary. (1 acre = 0.40 ha)

    Environmental organizations have yielded a range of estimates of the amount of forest clearing that is occurring in the Brazilian Amazon. Convert the first two estimates into hectares per day and compare the three estimates

- Estimate 1: 15 acre per minute
- Estimate 2: 22,000 acre per day
- Estimate 3: 8,000 ha per day

Review Key Terms

Match the Key Terms on the left with the definitions on the right.

_____1. Ecosystem services

_____2. Environmental indicator

_____3. Biodiversity

_____4. Genetic diversity

_____5. Species

_____6. Species diversity

_____7. Speciation

_____8. Background extinction rate

_____9. Greenhouse gasses

_____10. Anthropogenic

_____11. Development

_____12. Sustainability

_____13. Sustainable development

_____14. Biophilia

_____15. Ecological footprint

a. The diversity of life forms in an environment

b. A measure of the genetic variation among individuals in a population

c. Development that balances current human well-being and economic advancement with resource management for the benefit of future generations

d. Love of life

e. Improvement in human well-being through economic advancement

f. An indicator that describes the current state of an environmental system

g. A group of organisms that is distinct from other groups in its morphology, behavior, or biochemical properties

h. Living on Earth in a way that allows humans to use its resources without depriving future generations of those resources

i. A measure of how much an individual consumes, expressed in an area of land

j. The evolution of new species

k. The number of species in a region or in a particular type of habitat

l. Derived from human activities

m. The average rate at which species become extinct over the long term

n. Gasses in Earth's atmosphere that trap heat near the surface

o. The processes by which life-supporting resources such as clean water, timber, fisheries, and agricultural crops are produced

Module 3: Scientific Method

BEFORE YOU READ THE MODULE

Focus on Learning Objectives

Use the module learning objectives to guide your reading. On a separate piece of paper, write down each objective and take notes to help you meet each learning objective. After studying this module, you should be able to:

- explain the scientific method and its application to the study of environmental problems.
- describe some of the unique challenges and limitations of environmental science.

Preview Key Terms

In a notebook or on a separate sheet of paper, create a table like the one shown here to help with learning new key terms in the module. Before you read, fill out the "Prediction" column. Write what you think the term might mean or what it makes you think about. Use examples from your everyday life. There are no wrong answers!

Key Term	Prediction	Definition
Write key term here.	*Write what you think the term means in this column.*	*Define the term here. Add an example and use it in a sentence.*

Key Terms

Scientific method   Sample size (*n*)   Theory
Hypothesis   Accuracy   Control group
Null hypothesis   Precision   Natural experiment
Replication   Uncertainty

WHILE YOU READ THE MODULE

Define Key Terms

When you come across a new key term while reading the module, copy the definition into the "Definition" column of your key terms table. Add an example and use the term in a sentence. Compare your initial ideas to the actual definition.

Study the Figure

Examine Figure 3.1, "The scientific method" on page 19. Complete this question in your notebook or on a separate sheet of paper.

1. An environmental science student conducting field observations noted dissolved oxygen levels in sunny areas of a lake were lower compared to dissolved oxygen levels recorded from shady areas of a lake. Use the flow chart in Figure 3.1 as a model to design an experiment to determine how temperature affects dissolved oxygen levels in an aquatic system. Include the following steps in your design:

   - Observation and questioning
   - Testable hypothesis
   - Data collection procedure, including an experimental group and a control group
   - Analysis of data, including a statement describing the importance of repeated trials (You may create sample data or simply describe the type of data to be collected.)

AFTER YOU READ THE MODULE

Review Key Terms

Match the key terms on the left with the definitions on the right.

_____1. Scientific method

_____2. Hypothesis

_____3. Null hypothesis

_____4. Replication

_____5. Sample size

_____6. Accuracy

_____7. Precision

_____8. Uncertainty

_____9. Theory

a. A statement or idea that can be falsified, or proven wrong

b. An objective method to explore the natural world, draw inferences from it, and predict the outcome of certain events, processes, or changes

c. An estimate of how much a measured or calculated value differs from a true value

d. The data collection procedure of taking repeated measurements

e. A natural event that acts as an experimental treatment in an ecosystem

f. A hypothesis that has been repeatedly tested and confirmed by multiple groups of researchers and has reached wide acceptance

g. In a scientific investigation, a group that experiences exactly the same conditions as the experiment group, except for the single variable under study

h. How close a measured value is to the actual or true value

i. The number of times a measurement is replicated in the data collection process

_____10. Control group

j. A testable theory or supposition about how something works

_____11. Natural experiment

k. How close the repeated measurements of a sample are to one another

Chapter (1) Review Exercises

Check Your Understanding

Review "Learning Objectives Revisited" on page 27 of your textbook. Compare the notes you took while reading each module. Complete these exercises to review the chapter.

1. What disciplines are incorporated into the study of environmental science?

2. List the 5 key global-scale environmental indicators.

3. Describe the following: genetic diversity, species diversity, and ecosystem diversity.

4. Give an example of an anthropogenic activity.

5. Currently, what is the size of the human population?

6. What is a person's ecological footprint?

7. List the steps in the scientific method.

Practice for Free-Response Questions

Complete this exercise to build and practice the skills you will need to answer free-response questions on the exam. Use a separate sheet of paper if necessary.

Humans manipulate the environment more than any other species. Complete the table below to identify how human activities have affected the environment and to identify relevant environmental indicators that can help us evaluate the current state of the system.

Human Activity	Environmental Impact	Environmental Indicator
Increased numbers of human population		
Land use changes/ increased urbanization/ agriculture		
Increased rate of species extinctions		
Food production		
Burning fossil fuels		
Overfishing		

Review and Reflect

Complete these activities to solidify your knowledge of the chapter concepts and key terms. Use a notebook or a separate sheet of paper if necessary.

1. Review your key terms table for each module.

   (a) Which words did you already know? Which were new to you?
   (b) Write a new sentence using each key term.
   (c) Create a set of flash cards that includes each key term. Use the cards to review terms that were new or challenging.
   (d) When you feel comfortable with the new or challenging terms, review all of the cards, including those with familiar terms.
   (e) Save your cards to review before an exam.

2. What are the challenging concepts from this chapter?

   (a) Identify any concepts you found particularly challenging in this chapter.
   (b) Create a list of topics you need to review in preparation for an exam.

3. What questions do you have about concepts in the chapter?

   (a) Note any further questions you might have about material in the chapter.
   (b) Work with a partner to discuss these questions and ask your teacher for help as needed.

4. Write five possible multiple-choice questions based on this chapter. Work with a partner to quiz each other in preparation for an exam.

Chapter ② Environmental Systems

Chapter Summary

This chapter reviews the concept of systems in nature, the different properties of matter, and the various forms of energy. It offers an overview of key ideas from chemistry and physics that scientists use to understand the environment and to measure human impact on the environment. The chapter consists of the following 2 modules:

- **Module 4:** Systems and Matter
- **Module 5:** Energy, Flows, and Feedbacks

Chapter Opening Case: *A Lake of Salt Water, Dust Storms, and Endangered Species*

This opening case shows us that the activities of humans, the lives of other organisms, and abiotic processes in the environment are interconnected. Humans, water, animals, plants, and the desert environment all interact at Mono Lake to create a complex environmental system. The story also demonstrates how a single change made to an ecosystem often has wide-ranging and unexpected consequences.

Do the Math

This chapter contains the following "Do the Math" feature to help prepare you for calculation questions you might encounter on the exam.

- "Calculating Energy Use and Converting Units" (page 46)

To make sure you understand the concepts and techniques presented in these boxes, do the practice problems presented in the text as well as the additional "Practice the Math" problems that appear in Module 5 of this study guide.

Module 4: Systems and Matter

BEFORE YOU READ THE MODULE

Focus on Learning Objectives

Use the module learning objectives to guide your reading. On a separate piece of paper, write down each objective and take notes to help you meet each learning objective. After studying this module, you should be able to:

- describe how matter comprises atoms and molecules that move among different systems.
- explain why water is an important component of most environmental systems.
- discuss how matter is conserved in chemical and biological systems.

Preview Key Terms

In a notebook or on a separate sheet of paper, create a table like the one shown here to help with learning new key terms in the module. Before you read, fill out the "Prediction" column. Write what you think the term might mean or what it makes you think about. Use examples from your everyday life. There are no wrong answers!

Key Term	Prediction	Definition
Write key term here.	*Write what you think the term means in this column.*	*Define the term here. Add an example and use it in a sentence.*

Key Terms

Matter
Mass
Atom
Element
Periodic table
Molecules
Compounds
Atomic number
Mass number
Isotopes
Radioactive decay
Half-life

Covalent bond
Ionic bond
Hydrogen bond
Polar molecule
Surface tension
Capillary action
Acid
Base
pH
Chemical reaction
Law of conservation of matter

Inorganic compounds
Organic compounds
Carbohydrates
Proteins
Nucleic acids
DNA (deoxyribonucleic acid)
RNA (ribonucleic acid)
Lipids
Cell

WHILE YOU READ THE MODULE

Define Key Terms

When you come across a new key term while reading the module, copy the definition into the "Definition" column of your key terms table. Add an example and use the term in a sentence. Compare your initial ideas to the actual definition.

Study the Figure

Examine Figure 4.7, "Surface tension" on page 39. The pH scale is logarithmic, meaning that each number on the scale changes by a factor of ten. So a substance that has a pH of 5 is 10×10, or 100 times more acidic than a substance with a pH of 7.

1. Based on the pH scale in Figure 4.7, how much more acidic are lakes affected by acid rain than sea water?

AFTER YOU READ THE MODULE

Review Key Terms

Match the key terms on the left with the definitions on the right.

_____1. Matter	a. A chemical bond between two ions of opposite charges.
_____2. Mass	b. A substance that contributes hydroxide ions to a solution
_____3. Atom	c. The number of protons in the nucleus of a particular element
_____4. Element	d. The smallest particle that can contain the chemical properties of an element
_____5. Periodic table	e. A property of water that results from the cohesion of water molecules at the surface of a body of water and that creates a sort of skin on the water's surface
_____6. Molecule	f. The number that indicates the relative strength of acids and bases in a substance
_____7. Compound	g. The spontaneous release of material from the nucleus of radioactive isotopes
_____8. Atomic number	h. A substance that contributes hydrogen ions to a solution
_____9. Mass number	i. A substance composed of atoms that cannot be broken down into smaller, simpler components

_____10. Isotopes

j. A law of nature stating that matter cannot be created or destroyed; it can only change form.

_____11. Radioactive decay

k. A measurement of the total number of protons and neutrons in an element

_____12. Half-life

l. A nucleic acid, the genetic material that contains the code for reproducing the components of the next generation, and which organisms pass on to their offspring

_____13. Covalent bond

m. Anything that occupies space and has mass

_____14. Ionic bond

n. A compound that contains carbon-carbon and carbon-hydrogen bonds

_____15. Hydrogen bond

o. A measurement of the amount of matter an object contains

_____16. Polar molecule

p. A compound that does not contain the element carbon or contains carbon bound to elements other than hydrogen

_____17. Surface tension

q. A particle containing more than one atom

_____18. Capillary action

r. A chart of all chemical elements currently known, organized by their properties

_____19. Acid

s. A highly organized living entity that consists of the four types of macromolecules and other substances in a watery solution, surrounded by a membrane

_____20. Base

t. The bond formed when elements share electrons

_____21. pH

u. a smaller organic biological molecules that does not mix with water

_____22. Chemical reaction

v. A molecule in which one side is more positive and the other side is more negative

_____23. Law of conservation of matter

w. A critical component of living organisms made up of a long chain of nitrogen-containing organic molecules known as amino acids

_____24. Inorganic compound

x. A nucleic acid that translates the code stored in DNA, which makes possible the synthesis of proteins

_____25. Organic compound

y. Atoms of the same element with different numbers of neutrons

_____26. Carbohydrate

z. A weak chemical bond that forms when hydrogen atoms that are covalently bonded to one atom are attracted to another atom on another molecule

_____27. Protein

_____28. Nucleic acid

_____29. DNA (deoxyribonucleic acid)

_____30. RNA (ribonucleic acid)

_____31. Lipid

_____32. Cell

aa. A property of water that occurs when adhesion of water molecules to a surface is stronger than cohesion between the molecules

bb. Organic compounds found in all living cells

cc. The time it takes for one-half of an original radioactive parent atom to decay

dd. A molecule containing more than one element

ee. A reaction that occurs when atoms separate from molecules or recombine with other molecules

ff. A compound composed of carbon, hydrogen, and oxygen atoms

Module 5: Energy, Flows, and Feedbacks

BEFORE YOU READ THE MODULE

Focus on Learning Objectives

Use the module learning objectives to guide your reading. On a separate piece of paper, write down each objective and take notes to help you meet each learning objective. After studying this module, you should be able to:

- distinguish among various forms of energy and understand how they are measured.
- discuss the first and second laws of thermodynamics and explain how they influence environmental systems.
- explain how scientists keep track of energy and matter inputs, outputs, and changes to environmental systems.

Preview Key Terms

In a notebook or on a separate sheet of paper, create a table like the one shown here to help with learning new key terms in the module. Before you read, fill out the "Prediction" column. Write what you think the term might mean or what it makes you think about. Use examples from your everyday life. There are no wrong answers!

Key Term	Prediction	Definition
Write key term here.	*Write what you think the term means in this column.*	*Define the term here. Add an example and use it in a sentence.*

Key Terms

Energy
Electromagnetic radiation
Photons
Potential energy
Kinetic energy
Chemical energy
Joule
Power

Temperature
First law of thermodynamics
Second law of thermodynamics
Energy efficiency
Energy quality
Entropy
Open system
Closed system

Inputs
Outputs
Systems analysis
Steady state
Negative feedback loop
Positive feedback loop

WHILE YOU READ THE MODULE

Define Key Terms

When you come across a new key term while reading the module, copy the definition into the "Definition" column of your key terms table. Add an example and use the term in a sentence. Compare your initial ideas to the actual definition.

Practice the Math: Calculating Energy Use and Converting Units

Read "Do the Math: Calculating Energy Use and Converting Units" on page 46. Try "Your Turn." For more math practice, do the following problems. Remember to show your work.

1. A family wants to calculate the cost of doing their laundry. The washer uses 1500 watts and the drier uses 2,000 watts. The family pays $0.10 per kilowatt-hour. Each appliance runs approximately 30 minutes each day. How much does the family spend per week to run these appliances?

2. An older model dishwasher uses 1400 watts when the motor is running. The newer more efficient dishwasher uses 600 watts. Assume the dishwasher is used 3 times a week and runs 3 hours each time it is used.

   (a) How much energy in kilowatt-hours per year will you save by using the more efficient dishwasher?

   (b) Assume you are paying $0.10 per kilowatt-hour for electricity. A new dishwasher will cost $600. You will receive a $100 rebate from the electric company for purchasing a more energy-efficient dishwasher. How long will it be before your energy savings compensate you for the cost of the new appliance?

Study the Figure

Use Figure 5.6, "The second law of thermodynamics" on page 49 to answer the following question.

1. Incandescent light bulbs convert 95 percent of supplied energy to heat rather than light. Calculate the conversion of coal into light if a compact fluorescent bulb with an efficiency rating of 70 percent is substituted for the incandescent bulb shown in the figure.

Review Key Terms

Match the key terms on the left with the definitions on the right.

_____1. Energy

a. The measure of the average kinetic energy of a substance

_____2. Joule

b. A physical law which states that energy can neither be created nor destroyed but can change from one form to another

_____3. Power

c. The ease with which an energy source can be used for work

_____4. Electromagnetic radiation

d. The ratio of the amount of energy expended in the form you want to the total amount of energy that is introduced into the system

_____5. Photon

e. The amount of energy used when a 1-watt electrical device is turned on for 1 second

_____6. Potential energy

f. Stored energy that has not been released

_____7. Chemical energy

g. Randomness in a system

_____8. Kinetic energy

h. A system in which matter and energy exchanges do not occur across boundaries

_____9. Temperature

i. Potential energy stored in chemical bonds.

_____10. First law of thermodynamics

j. An addition to a system

_____11. Second law of thermodynamics

k. The ability to do work or transfer heat

_____12. Energy efficiency

l. An analysis to determine inputs, outputs, and changes in a system under various conditions

_____13. Energy quality

m. A massless packet of energy that carries electromagnetic radiation at the speed of light.

_____14. Entropy

n. A feedback loop in which a system responds to a change by returning to its original state, or by decreasing the rate at which the change is occurring

_____15. Open system

o. A state in which inputs equal outputs, so that the system is not changing over time

_____16. Closed system

p. The rate at which work is done

_____17. Input

q. The energy of motion

_____18. Output

r. A feedback loop in which change in a system is amplified

_____19. Systems analysis

_____20. Steady state

_____21. Negative feedback loop

_____22. Positive feedback loop

s. A loss from a system

t. A system in which exchanges of matter or energy occur across system boundaries

u. The physical law stating that when energy is transformed, the quantity of energy remains the same, but its ability to do work diminishes

v. A form of energy emitted by the Sun that includes, but is not limited to, visible light, ultraviolet light, and infrared energy

Chapter (2) Review Exercises

Check Your Understanding

Review "Learning Objectives Revisited" on page 57 of your textbook. Compare the notes you took while reading each module. Complete these exercises to review the chapter.

1. Briefly explain radioactive decay.

2. What is an element's half-life?

3. List some properties of water and describe how these properties are important to living systems.

4. If a substance has a pH of 3, how many more times acidic is it than a substance with a pH of 5?

5. Give an example of potential energy and of kinetic energy.

6. Explain the first and second laws of thermodynamics in your own words.

7. What is the energy efficiency of a coal burning power plant, an incandescent light bulb and the electrical transmission lines between the power plant and the house?

8. Give an example of a negative feedback loop and a positive feedback loop.

Practice for Free-Response Questions

Complete this exercise to build and practice the skills you will need to answer free-response questions on the exam. Use a separate sheet of paper if necessary.

Explain how energy is transformed from the sun into biologically usable energy. Include the first and second laws of thermodynamics in your discussion.

Review and Reflect

Complete these activities to solidify your knowledge of the chapter concepts and key terms. Use a notebook or a separate sheet of paper if necessary.

1. Review your key terms table for each module.

   (a) Which words did you already know? Which were new you?
   (b) Write a new sentence using each key term.
   (c) Create a set of flash cards that includes each key term. Use the cards to review terms that were new or challenging.
   (d) When you feel comfortable with the new or challenging terms, review all of the cards, including those with familiar terms.
   (e) Save your cards to review before an exam.

2. What are the challenging concepts from this chapter?

   (a) Identify any concepts you found particularly challenging in this chapter.
   (b) Create a list of topics you need to review in preparation for an exam.

3. What questions do you have about concepts in the chapter?

   (a) Note any further questions you might have about material in the chapter.
   (b) Work with a partner to discuss these questions and ask your teacher for help as needed.

4. Write five possible multiple-choice questions based on this chapter. Work with a partner to quiz each other in preparation for an exam.

Unit 1 Multiple-Choice Review Exam

Choose the best answer.

1. Which is a biotic component of an ecosystem?
   (A) Soil
   (B) Water
   (C) Nitrogen
   (D) Sunflower
   (E) Rock

2. As a solid or liquid, water has its lowest density at
   (A) 0° Celsius.
   (B) 32° Celsius.
   (C) 100° Fahrenheit.
   (D) 4° Celsius.
   (E) 100° Celsius.

3. Earth is regulated by feedback loops that are
   (A) positive.
   (B) negative.
   (C) both positive and negative
   (D) getting weaker.
   (E) in a steady state.

4. How many acres are in 7 square miles? (1 square mile = 640 acres)
   (A) 0.4480
   (B) 4.480
   (C) 44.80
   (D) 448.0
   (E) 4,480

5. Which of the following is NOT an example of an anthropogenic activity?
   (A) Burning fossil fuels for electricity
   (B) Burning fossil fuels in vehicles
   (C) Volcanic eruptions
   (D) Deforestation for planting crops
   (E) Over-harvesting of ocean resources

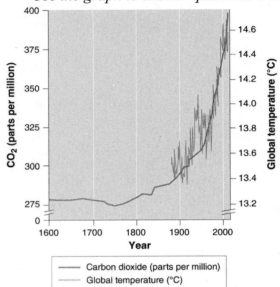

Figure 2.5
Environmental Science for AP®, Second Edition
Data from http://data.giss.nasa.gov/gistemp/graphs_v3/ and http://www.esrl.noaa.gov/gmd/ccgg/trends/#mlo_full

6. When carbon dioxide levels were at 325 ppm, what was the approximate global temperature in degrees Celsius?
   (A) 13.2
   (B) 14.1
   (C) 13.5
   (D) 13.8
   (E) 14.8

7. If carbon dioxide levels hit 400 ppm, what do you estimate global temperature to be?
   (A) 14.8° Celsius
   (B) 13.8° Celsius
   (C) 17.2° Celsius
   (D) 2,100° Celsius
   (E) 2,500° Celsius

8. If the trend in global surface temperatures continues, what year do you estimate temperatures to become 15.0° Celsius?
   (A) 2000
   (B) 2100
   (C) 2500
   (D) 3000
   (E) 3500

9. Greenhouse gases in the atmosphere help
   (A) keep UV light from reaching Earth.
   (B) regulate temperatures near Earth's surface.
   (C) heat to be released back to space.
   (D) keep the ozone layer intact.
   (E) Earth stay cooler.

10. If population of the United States is 307,000,000 and approximately 50 billion bottles of water are consumed in the United States each year, how many bottles are consumed per capita?
   (A) 1
   (B) 7
   (C) 50
   (D) 100
   (E) 160

11. If there are 364,000 infants born every day and 152,000 people die each day, the net result is 212,000 new inhabitants on Earth each day. What is the approximate net gain of new inhabitants on Earth in one year?
   (A) 1 million
   (B) 7 million
   (C) 9.6 billion
   (D) 10.5 billion
   (E) 77 million

12. A hypothesis is
   (A) testable.
   (B) determined from data.
   (C) an observation.
   (D) a scientific theory.
   (E) an experimental control.

Match the following words to their correct definition

13. _____Accuracy   (A) How close the repeated measurements of a sample are to one
14. _____Precision   another
15. _____Uncertainty (B) An estimate of how much a measured or calculated value differs
                               from a true value
                          (C) How close a measured value is to the actual or true value

16. An activist group is protesting a landfill that is being built in a poor neighborhood. The group objects that people of lower socioeconomic status do not have enough influence to object to the landfill. This group is concerned with
    (A) social action.
    (B) environmentalism.
    (C) environmental justice.
    (D) The three R's.
    (E) environmental economics.

17. What type of chemical reaction occurs when one atom loses an electron and another atom gains the electron?
    (A) Covalent bond
    (B) Ionic bond
    (C) Hydrogen bond
    (D) Polar bond
    (E) Energy bond

18. Household bleach with a pH of 13 is how many times more basic than pure water with a pH of 7?
    (A) 100
    (B) 1,000
    (C) 6
    (D) 6,000
    (E) 1,000,000

19. Energy in the form of coal is an example of
    I. potential energy.
    II. kinetic energy.
    III. renewable energy.

    (A) I only
    (B) II only
    (C) III only
    (D) I and II
    (E) I, II, and III

20. The conversion of food into energy in the form of body heat demonstrates
    (A) potential energy.
    (B) kinetic energy.
    (C) energy of motion.
    (D) the first law of thermodynamics.
    (E) the second law of thermodynamics.

21. If energy was not constrained under the second law of thermodynamics how efficient would our generation of energy be?
    (A) 10 percent
    (B) 50 percent
    (C) 75 percent
    (D) 100 percent
    (E) 0 percent

22. Which can be described as a system responding to change by returning to its original state or by decreasing the rate at which change is occurring?
    (A) Adaptive management
    (B) A steady state
    (C) A negative feedback loop
    (D) Entropy
    (E) A positive feedback loop

23. According to the laws of thermodynamics
    I.   energy is neither created nor destroyed.
    II. when energy is transformed, the quantity of energy remains the same.
    III. when energy is transformed, its ability to do work diminishes.

    (A) I only
    (B) II only
    (C) III only
    (D) I and III only
    (E) I, II, and III

24. Most energy on Earth comes from
    (A) plants.
    (B) hydroelectric power.
    (C) geothermal sources.
    (D) the Sun.
    (E) entropy.

25. Changes in climate are often difficult to predict due to feedback loops that regulate Earth's temperature. Which describes a negative feedback loop?

    I. As surface temperatures increase, permafrost thaws which releases additional methane to the atmosphere.

    II. Increases in atmospheric water causes more clouds to form which in turn increases the amount of solar radiation reflected away from Earth's surface.

    III. Increased carbon dioxide provides raw materials for photosynthetic organisms which leads to increased plant growth. Greater photosynthetic activity leads to less atmospheric carbon.

(A) I only

(B) II only

(C) III only

(D) I and II

(E) II and III

26. Which does NOT accurately describe the process of speciation?

(A) Speciation occurs at a rate of 1-3 species per year.

(B) Speciation increases species diversity within a region.

(C) Speciation may arise from changes in the environment.

(D) Speciation rates currently exceed extinction rates.

(E) Speciation results in a distinct group of organisms that breed and produce fertile offspring.

27. The scientific method provides an objective and methodical method to evaluate environmental phenomenon. Which component of data collection improves the validity of the results?

(A) Replication

(B) A null hypothesis

(C) Uncertainty

(D) Accuracy

(E) Precision

28. Sustainable development includes which of the following key components?

    I. Current human well-being

    II. Economic growth

    III. Resource management that will benefit future generations.

(A) I and II

(B) I and IIII

(C) I, II, and III

(D) II and III

(E) III only

Chapter (3) Ecosystem Ecology

Chapter Summary

This chapter explains how ecosystems control the movement of the water, energy, and nutrients that organisms must have to grow and reproduce. Key concepts include food webs, the water (hydrologic) cycle, and the cycling of the macronutrients carbon, nitrogen, phosphorous, and sulfur. The AP® exam frequently contains numerous multiple-choice questions about the cycles, and the biogeochemical cycles have shown up on free-response questions. Make sure you understand how nutrients cycle, the parts and names of each cycle, and the phases of matter in which each occurs. The chapter consists of 3 modules:

- **Module 6:** The Movement of Energy
- **Module 7:** The Movement of Matter
- **Module 8:** Responses to Disturbances

Chapter Opening Case: *Reversing the Deforestation of Haiti*

This opening case shows us how economics is being used to solve an environmental problem. This case shows that keeping a mango tree that can provide between $70 and $150 each year is better than cutting down a tree and burning it. Many problems can be solved by using economic principles.

Do the Math

This chapter contains the following "Do the Math" box to help prepare you for calculation questions you might encounter on the exam.

- "Raising Mangoes" (page 81)

To make sure you understand the concepts and techniques presented in this box, do the practice problems presented in the text as well as the additional "Practice the Math" problems that appear in Module 7 of this study guide

Module 6: The Movement of Energy

BEFORE YOU READ THE MODULE

Focus on Learning Objectives

Use the module learning objectives to guide your reading. On a separate piece of paper, write down each objective and take notes to help you meet each learning objective. After studying this module, you should be able to:

- explain the concept of ecosystem boundaries.
- describe the processes of photosynthesis and respiration.
- distinguish among the trophic levels that exist in food chains and food webs.
- quantify ecosystem productivity.
- explain energy transfer efficiency and trophic pyramids.

Preview Key Terms

In a notebook or on a separate sheet of paper, create a table like the one shown here to help with learning new key terms in the module. Before you read, fill out the "Prediction" column. Write what you think the term might mean or what it makes you think about. Use examples from your everyday life. There are no wrong answers!

Key Term	Prediction	Definition
Write key term here.	*Write what you think the term means in this column.*	*Define the term here. Add an example and use it in a sentence.*

Key Terms

Biosphere
Producer
Autotrophs
Photosynthesis
Cellular respiration
Aerobic respiration
Anaerobic respiration
Consumer
Heterotroph

Herbivore
Primary consumer
Carnivore
Secondary consumer
Tertiary consumer
Trophic level
Food chain
Food web
Scavenger

Detritivore
Decomposer
Gross primary productivity (GPP)
Net primary productivity (NPP)
Biomass
Standing crop
Ecological efficiency
Trophic pyramid

WHILE YOU READ THE MODULE

Define Key Terms

When you come across a new key term while reading the module, copy the definition into the "Definition" column of your key terms table. Add an example and use the term in a sentence. Compare your initial ideas to the actual definition

AFTER YOU READ THE MODULE

Review Key Terms

Match the key terms on the left with the definitions on the right.

_____1. Biosphere

a. An organism that is incapable of photosynthesis and must obtain its energy by consuming other organisms

_____2. Producer (autotroph)

b. The successive levels of organisms consuming one another

_____3. Photosynthesis

c. The region of our planet where life resides, the combination of all ecosystems on Earth

_____4. Cellular respiration

d. the energy captured by producers in an ecosystem minus the energy producers respire

_____5. Aerobic respiration

e. The sequence of consumption from producers through tertiary consumers

_____6. Anaerobic respiration

f. The process by which cells convert glucose into energy in the absence of oxygen

_____7. Consumer (heterotroph)

g. An organism that uses the energy of the Sun to produce usable forms of energy

_____8. Herbivore (primary consumer)

h. A carnivore that eats secondary consumers

_____9. Carnivore

i. An organism that specializes in breaking down dead tissues and waste products into smaller particles

_____10. Secondary consumer

j. The total mass of all living matter in a specific area

_____11. Tertiary consumer

k. A representation of the distribution of biomass, numbers, or energy among trophic levels

_____12. Trophic levels

l. The proportion of consumed energy that can be passed from one trophic level to another

_____13. Food chain

m. The total amount of solar energy that producers in an ecosystem capture via photosynthesis over a given amount of time

_____14. Food web

n. The process by which producers use solar energy to convert carbon dioxide and water into glucose

_____15. Scavenger

o. The amount of biomass present in an ecosystem at a particular time

_____16. Detritivore

p. The process by which cells unlock the energy of chemical compounds

_____17. Decomposers

q. A consumer that eats producers

_____18. Gross primary productivity (GPP)

r. A carnivore that eats primary consumers

_____19. Net primary productivity (NPP)

s. Fungi and bacteria that convert organic matter into small elements and molecules that can be recycled back into the ecosystem

_____20. Biomass

t. An organism that consumes dead animals

_____21. Standing crop

u. A complex model of how energy and matter move between trophic levels

_____22. Ecological efficiency

v. A consumer that eats other consumers

_____23. Trophic pyramid

w. The process by which cells convert glucose and oxygen into energy, carbon dioxide, and water

Module 7: The Movement of Matter

BEFORE YOU READ THE MODULE

Focus on Learning Objectives

Use the module learning objectives to guide your reading. On a separate piece of paper, write down each objective and take notes to help you meet each learning objective. After studying this module, you should be able to:

- Describe how water cycles within ecosystems.
- Explain how carbon cycles within ecosystems.
- Describe how nitrogen cycles within ecosystems.
- Explain how phosphorus cycles within ecosystems.
- Discuss the movement of calcium, magnesium, potassium, and sulfur within ecosystems

Preview Key Terms

In a notebook or on a separate sheet of paper, create a table like the one shown here to help with learning new key terms in the module. Before you read, fill out the "Prediction" column. Write what you think the term might mean or what it makes you think about. Use examples from your everyday life. There are no wrong answers!

Key Term	Prediction	Definition
Write key term here.	*Write what you think the term means in this column.*	*Define the term here. Add an example and use it in a sentence.*

Key Terms

Biogeochemical cycle
Macronutrient
Hydrologic cycle
Transpiration
Evapotranspiration
Runoff
Carbon cycle

Limiting nutrient
Nitrogen cycle
Nitrogen fixation
Nitrification
Assimilation
Mineralization
Ammonification

Denitrification
Leaching
Phosphorus cycle
Algal bloom
Hypoxic
Sulfur cycle

WHILE YOU READ THE MODULE

Define Key Terms

When you come across a new key term while reading the module, copy the definition into the "Definition" column of your key terms table. Add an example and use the term in a sentence. Compare your initial ideas to the actual definition.

Study the Figure

Use Figure 7.1, "The hydrologic cycle" on page 80 to answer the following question.

1. Choose two processes of the hydrologic cycle and describe how humans have affected them.

Practice the Math: Raising Mangoes

Read "Do the Math: Raising Mangoes." Try "Your Turn," For more math practice, do the following problems. Remember to show your work. Use a separate piece of paper if necessary.

1. A farmer has decided to plant and sell Christmas trees. The farmer estimates that each tree will sell for $75.

    (a) How much will the farmer earn on sales of 300 trees?

    (b) If each tree take 2 gallons of water per day to thrive, how many gallons of water will it take to grow the trees each year?

    (c) If a new sapling costs $1.25 and the farmer wants to replant the farm at the end of the year with 300 new trees, how much will the farmer need to invest?

2. Banana saplings cost $8 each. Once the trees mature, each tree will produce $50 worth of fruit per year. A village of 300 people decides to pool its resources to establish a community banana plantation. Their goal is to generate an income of $200 per year for each person in the village.

    (a) How many mature trees will the village need to meet the goal?

    (b) Each tree requires 10 m^2 of space. How many hectares must the village set aside for the plantation? (1 m^2 = 0.0001 ha)

    (c) Each tree requires 15 L of water to be pumped in every day during the 6 hot months of the year (180 days) and no additional water during the other six months. How many liters of water are needed to be pumped in each year?

Study the Figure

Use Figure 7.2, "The carbon cycle" on page 83 to answer the following question.

1. Carbon may be cycled through various biological and geological cycles. Categorize the processes represented in the carbon cycle as a biotic or an abiotic process. Determine if each process retains or releases carbon.

Biological	Geological

Study the Figure

Use Figure 7.3, "The nitrogen cycle" on page 84 to answer the following question.

1. Examine the nitrogen cycle. List each process within the nitrogen cycle and determine its products.

Process	Product

AFTER YOU READ THE MODULE

Review Key Terms

Match the key terms on the left with the definitions on the right.

_____1. Biogeochemical cycle

a. The conversion of ammonia (NH_4^+) into nitrite (NO_2^-) and then into nitrate (NO_3^-)

_____2. Hydrologic cycle

b. The conversion of nitrate (NO_3^-) in a series of steps into the gases nitrous oxide (N_2O) and, eventually, nitrogen gas (N_2), which is emitted into the atmosphere

_____3. Transpiration

c. A rapid increase in the algal population of a waterway

_____4. Evapotranspiration

d. The movement of phosphorus around the biosphere

_____5. Runoff

_____6. Carbon cycle

_____7. Macronutrient

_____8. Limiting nutrient

_____9. Nitrogen cycle

_____10. Nitrogen fixation

_____11. Nitrification

_____12. Assimilation

_____13. Mineralization

_____14. Ammonification

_____15. Denitrification

_____16. Leaching

_____17. Phosphorus cycle

_____18. Algal bloom

_____19. Hypoxic

_____20. Sulfur cycle

e. The process by which producers incorporate elements into their tissues

f. Water that moves across the land surface and into streams and rivers

g. The movement of carbon around the biosphere

h. The movement of nitrogen around the biosphere.

i. The movement of water through the biosphere

j. A nutrient required for the growth of an organism but available in a lower quantity than other nutrients.

k. The process by which fungal and bacterial decomposers break down the organic matter found in dead bodies and waste products and convert it into inorganic compounds

l. The release of water from leaves during photosynthesis

m. The process by which fungal and bacterial decomposers break down the organic nitrogen found in dead bodies and waste products and convert it into inorganic ammonium (NH_4^+)

n. The movement of sulfur around the biosphere

o. Low in oxygen

p. The movements of matter within and between ecosystems

q. The transportation of dissolved molecules through the soil via groundwater

r. The combined amount of evaporation and transpiration

s. A process by which some organisms can convert nitrogen gas molecules directly into ammonia

t. One of six key elements that organisms need in relatively large amounts: nitrogen, phosphorus, potassium, calcium, magnesium, and sulfur

Module 8: Responses to Disturbances

BEFORE YOU READ THE MODULE

Focus on Learning Objectives

Use the module learning objectives to guide your reading. On a separate piece of paper, write down each objective and take notes to help you meet each learning objective. After studying this module, you should be able to:

- explain the insights gained from watershed studies.
- distinguish between ecosystem resistance and ecosystem resilience.
- explain the intermediate disturbance hypothesis.

Preview Key Terms

In a notebook or on a separate sheet of paper, create a table like the one shown here to help with learning new key terms in the module. Before you read, fill out the "Prediction" column. Write what you think the term might mean or what it makes you think about. Use examples from your everyday life. There are no wrong answers!

Key Term	Prediction	Definition
Write key term here.	*Write what you think the term means in this column.*	*Define the term here. Add an example and use it in a sentence.*

Key Terms

Disturbance

Watershed

Resistance

Resilience

Restoration ecology

Intermediate disturbance hypothesis

WHILE YOU READ THE MODULE

Define Key Terms

When you come across a new key term while reading the module, copy the definition into the "Definition" column of your key terms table. Add an example and use the term in a sentence. Compare your initial ideas to the actual definition.

Review Key Terms

Match the key terms on the left with the definitions on the right.

_____1. Disturbance

a. A measure of how much a disturbance can affect flows of energy and matter in an ecosystem

_____2. Watershed

b. The hypothesis that ecosystems experiencing intermediate levels of disturbance are more diverse than those with high or low disturbance levels

_____3. Resistance

c. All land in a given landscape that drains into a particular stream, river, lake, or wetland

_____4. Resilience

d. The rate at which an ecosystem returns to its original state after a disturbance

_____5. Restoration ecology

e. An event, caused by physical, chemical, or biological agents, resulting in changes in population size or community composition

_____6. Intermediate disturbance hypothesis

f. The study and implementation of restoring damaged ecosystems

Chapter ③ Review Exercises

Check Your Understanding

Review "Learning Objectives Revisited" on page 97 of your textbook. Compare the notes you took while reading each module. Complete these exercises to review the chapter.

1. Write the formula for photosynthesis.

2. Write the formula for cellular respiration.

3. Examine Figure 6.9, which shows the amount of energy that is present at each trophic level in the Serengeti ecosystem. What does the amount of energy at each level tell us about ecological efficiency?

4. Explain the difference between the fast part of the carbon cycle and the slow part of the carbon cycle.

5. Briefly describe the steps of the nitrogen cycle.

Practice for Free-Response Questions

Complete this exercise to build and practice the skills you will need to answer free-response questions on the exam. Use a separate sheet of paper if necessary.

Matter and energy interact in ecosystems through biogeochemical cycles. These cycles are essential to ecosystem stability and also provide specific benefits to humans and other organisms. Identify an ecosystem service for each of the cycles or processes is the table.

System	Ecosystem Service
Biological energy transfer: Photosynthesis/ Cell respiration	
Decomposition	
Hydrologic cycle	
Carbon cycle	
Nitrogen cycle	
Phosphorus cycle	

Review and Reflect

Complete these activities to solidify your knowledge of the chapter concepts and key terms. Use a separate sheet of paper if necessary.

1. Review your key terms table for each module.

   (a) Which words did you already know? Which were new to you?
   (b) Write a new sentence using each key term.
   (c) Create a set of flash cards that includes each key term. Use the cards to review terms that were new or challenging.
   (d) When you feel comfortable with the new or challenging terms, review all of the cards, including those with familiar terms.
   (e) Save your cards to review before an exam.

2. What are the challenging concepts from this chapter?

   (a) Identify any concepts you found particularly challenging in this chapter.
   (b) Create a list of topics you need to review in preparation for an exam.

3. What questions do you have about concepts in the chapter?

   (a) Note any further questions you might have about material in the chapter.
   (b) Work with a partner to discuss these questions and ask your teacher for help as needed.

4. Write five possible multiple-choice questions based on this chapter. Work with a partner to quiz each other in preparation for an exam.

Chapter ④ Global Climates and Biomes

Chapter Summary

This chapter explores the factors that affect the distribution of heat and precipitation around the globe, including the unequal heating of Earth, air currents, and ocean currents. After looking at the pattern of climate differences around the world, the text describes how similar climates support similar types of plants and details characteristics of the nine terrestrial biomes. The chapter closes with a description of different aquatic biomes. The chapter consists of 5 modules:

- **Module 9:** The Unequal Heating of Earth
- **Module 10:** Air Currents
- **Module 11:** Ocean Currents
- **Module 12:** Terrestrial Biomes
- **Module 13:** Aquatic Biomes

Chapter Opening Case: *Growing Grapes to Make a Fine Wine*

This opening case illustrates that different regions of the world contain distinct climates and that these climates affect the species that can live in each region. Students learn that certain regions have comparable climates and therefore support similar plant and animal communities. The story of wine grapes further demonstrates that as climate changes, we can expect changes in the species that live in these regions as well as changes in the way humans use these ecosystems.

Module 9: The Unequal Heating of Earth

BEFORE YOU READ THE MODULE

Focus on Learning Objectives

Use the module learning objectives to guide your reading. On a separate piece of paper, write down each objective and take notes to help you meet each learning objective. After studying this module, you should be able to:

- identify the five layers of the atmosphere.
- discuss the factors that cause unequal heating of Earth.
- describe how Earth's tilt affects seasonal differences in temperatures.

Preview Key Terms

In a notebook or on a separate sheet of paper, create a table like the one shown here to help with learning new key terms in the module. Before you read, fill out the "Prediction" column. Write what you think the term might mean or what it makes you think about. Use examples from your everyday life. There are no wrong answers!

Key Term	Prediction	Definition
Write key term here.	*Write what you think the term means in this column.*	*Define the term here. Add an example and use it in a sentence.*

Key Terms

Climate
Weather
Troposphere

Stratosphere
Albedo

WHILE YOU READ THE MODULE

Define Key Terms

When you come across a new key term while reading the module, copy the definition into the "Definition" column of your key terms table. Add an example and use the term in a sentence. Compare your initial ideas to the actual definition.

AFTER YOU READ THE MODULE

Review Key Terms

Match the key terms on the left with the definitions on the right.

_____1. Climate

_____2. Weather

_____3. Troposphere

_____4. Stratosphere

_____5. Albedo

a. A layer of the atmosphere closest to the surface of Earth, extending up to approximately 16 km (10 miles)

b. The layer of the atmosphere above the troposphere, extending roughly 16 to 50 km (10–31 miles) above the surface of Earth

c. The average weather that occurs in a given region over a long period of time.

d. The percentage of incoming sunlight reflected from a surface

e. The short-term conditions of the atmosphere in a local area, which include temperature, humidity, clouds, precipitation, and wind speed

Module 10: Air Currents

BEFORE YOU READ THE MODULE

Focus on Learning Objectives

Use the module learning objectives to guide your reading. On a separate piece of paper, write down each objective and take notes to help you meet each learning objective. After studying this module, you should be able to:

- explain how the properties of air affect the way it moves in the atmosphere.
- identify the factors that drive atmospheric convection currents.
- describe how Earth's rotation affects the movement of air currents.
- explain how the movement of air currents over mountain ranges affects climates.

Preview Key Terms

In a notebook or on a separate sheet of paper, create a table like the one shown here to help with learning new key terms in the module. Before you read, fill out the "Prediction" column. Write what you think the term might mean or what it makes you think about. Use examples from your everyday life. There are no wrong answers!

Key Term	Prediction	Definition
Write key term here.	*Write what you think the term means in this column.*	*Define the term here. Add an example and use it in a sentence.*

Key Terms

Saturation point
Adiabatic cooling
Adiabatic heating
Latent heat release
Atmospheric convection currents
Hadley cell

Intertropical convergence zone (ITCZ)
Polar cell
Ferrell cell
Coriolis effect
Rain shadow

Define Key Terms

When you come across a new key term while reading the module, copy the definition into the "Definition" column of your key terms table. Add an example and use the term in a sentence. Compare your initial ideas to the actual definition.

Study the Figure

Use Figure 10.6, "Prevailing wind patterns" on page 115 to answer the following questions.

1. Explain how prevailing wind patterns are influenced by the combination of atmospheric convection currents and the Coriolis effect.

2. The prevailing winds are named after the direction from which they come. From memory, draw a model indicating the direction of the prevailing winds in each hemisphere of the earth. Label the northeast trade winds, the southeast trade winds, and the westerlies. Check your model using Figure 10.6.

Review Key Terms

Match the key terms on the left with the definitions on the right.

_____1. Saturation point

    a. Global patterns of air movement that are initiated by the unequal heating of Earth

_____2. Adiabatic cooling

    b. A convection current in the atmosphere that cycles between the equator and 30° N and 30° S

_____3. Adiabatic heating

    c. A convection current in the atmosphere, formed by air that rises at 60° N and 60° S and sinks at the poles, 90° N and 90° S

_____4. Latent heat release

    d. The cooling effect of reduced pressure on air as it rises higher in the atmosphere and expands

_____5. Atmospheric convection current

    e. The release of energy when water vapor in the atmosphere condenses into liquid water

_____6. Hadley cell

    f. The deflection of an object's path due to the rotation of Earth

_____7. Intertropical convergence zone (ITCZ)

    g. The maximum amount of water vapor in the air at a given temperature

_____8. Polar cell

    h. A convection current in the atmosphere that lies between Hadley cells and polar cells

_____9. Ferrell cell

    i. A region with dry conditions found on the leeward side of a mountain range as a result of humid winds from the ocean causing precipitation on the windward side

_____10. Coriolis effect

    j. The heating effect of increased pressure on air as it sinks toward the surface of Earth and decreases in volume

_____11. Rain shadow

    k. The latitude that receives the most intense sunlight, which causes the ascending branches of the two Hadley cells to converge

Module 11: Scientific Method

Focus on Learning Objectives

Use the module learning objectives to guide your reading. On a separate piece of paper, write down each objective and take notes to help you meet each learning objective. After studying this module, you should be able to:

- describe the patterns of surface ocean circulation.
- explain the mixing of surface and deep ocean waters from thermohaline circulation.
- identify the causes and consequences of the El Niño-Southern Oscillation.

Preview Key Terms

In a notebook or on a separate sheet of paper, create a table like the one shown here to help with learning new key terms in the module. Before you read, fill out the "Prediction" column. Write what you think the term might mean or what it makes you think about. Use examples from your everyday life. There are no wrong answers!

Key Term	Prediction	Definition
Write key term here.	*Write what you think the term means in this column.*	*Define the term here. Add an example and use it in a sentence.*

Key Terms

Gyres

Upwelling

Thermohaline circulation

El Niño–Southern Oscillation (ENSO)

WHILE YOU READ THE MODULE

Define Key Terms

When you come across a new key term while reading the module, copy the definition into the "Definition" column of your key terms table. Add an example and use the term in a sentence. Compare your initial ideas to the actual definition.

Study the Figure

Use Figure 11.2, "Thermohaline circulation" on page 119 to answer the following questions.

1. Which is more dense: cold water or warm water?

2. Which is more dense: salt water or fresh water?

3. Explain the mechanism that drives the thermohaline circulation.

4. Explain the environmental impacts of the thermohaline circulation.

AFTER YOU READ THE MODULE

Review Key Terms

Match the key terms on the left with the definitions on the right.

_____1. Gyre

a. The upward movement of ocean water toward the surface as a result of diverging currents

_____2. Upwelling

b. A large-scale pattern of water circulation that moves clockwise in the Northern Hemisphere and counterclockwise in the Southern Hemisphere

_____3. Thermohaline circulation

c. A reversal of wind and water currents in the South Pacific

_____4. El Niño–Southern Oscillation (ENSO)

d. An oceanic circulation pattern that drives the mixing of surface water and deep water

Module 12: Terrestrial Biomes

BEFORE YOU READ THE MODULE

Focus on Learning Objectives

Use the module learning objectives to guide your reading. On a separate piece of paper, write down each objective and take notes to help you meet each learning objective. After studying this module, you should be able to:

- explain how we define terrestrial biomes.
- interpret climate diagrams.
- identify the nine terrestrial biomes.

Preview Key Terms

In a notebook or on a separate sheet of paper, create a table like the one shown here to help with learning new key terms in the module. Before you read, fill out the "Prediction" column. Write what you think the term might mean or what it makes you think about. Use examples from your everyday life. There are no wrong answers!

Key Term	Prediction	Definition
Write key term here.	*Write what you think the term means in this column.*	*Define the term here. Add an example and use it in a sentence.*

Key Terms

Biomes
Tundra
Permafrost
Boreal forest
Temperate rainforest
Temperate seasonal forest

Woodland/shrubland
Temperate grassland / cold desert
Tropical rainforest
Tropical seasonal forest / savanna
Subtropical desert

WHILE YOU READ THE MODULE

Define Key Terms

When you come across a new key term while reading the module, copy the definition into the "Definition" column of your key terms table. Add an example and use the term in a sentence. Compare your initial ideas to the actual definition.

Study the Figure

Use Figure 12.4, "Climate diagrams" on page 124 to answer the following questions.

1. For each of the examples given in Figure 12.4, evaluate the temperature and precipitation patterns. Determine which biome each example most closely resembles based on the climatic patterns presented in the graphs. Support your answer.

2. Determine the average rainfall amounts and temperatures for each month in your region. On a separate piece of paper, construct a climatogram to represent the data you collect. Determine the identity of the biome you live in based on this data. Do the endemic plant species support your conclusions?

AFTER YOU READ THE MODULE

Review Key Terms

Match the key terms on the left with the definitions on the right.

_____1. Terrestrial biome

a. An impermeable, permanently frozen layer of soil

_____2. Aquatic biome

b. A coastal biome typified by moderate temperatures and high precipitation

_____3. Tundra

c. A warm and wet biome found between 20° N and 20° S of the equator, with little seasonal temperature variation and high precipitation

_____4. Permafrost

d. A biome characterized by hot, dry summers and mild, rainy winters

_____5. Boreal forest

e. An aquatic region characterized by a particular combination of salinity, depth, and water flow

_____6. Temperate rainforest

f. A cold and treeless biome with low-growing vegetation

_____7. Temperate seasonal forest

g. A forest biome made up primarily of coniferous evergreen trees that can tolerate cold winters and short growing seasons

_____8. Woodland/shrubland

    h. A biome prevailing at approximately 30° N and 30° S, with hot temperatures, extremely dry conditions, and sparse vegetation

_____9. Temperate grassland/cold desert

    i. A geographic region categorized by a particular combination of average annual temperature, annual precipitation, and distinctive plant growth forms on land

_____10. Tropical rainforest

    j. A biome marked by warm temperatures and distinct wet and dry seasons

_____11. Tropical seasonal forest/savanna

    k. A biome with warm summers and cold winters with over 1 m (39 inches) of precipitation annually

_____12. Subtropical desert

    l. A biome characterized by cold, harsh winters, and hot, dry summers

Module 13: Terrestrial Biomes

BEFORE YOU READ THE MODULE

Focus on Learning Objectives

Use the module learning objectives to guide your reading. On a separate piece of paper, write down each objective and take notes to help you meet each learning objective. After studying this module, you should be able to:

- identify the major freshwater biomes.
- identify the major marine biomes.

Preview Key Terms

In a notebook or on a separate sheet of paper, create a table like the one shown here to help with learning new key terms in the module. Before you read, fill out the "Prediction" column. Write what you think the term might mean or what it makes you think about. Use examples from your everyday life. There are no wrong answers!

Key Term	Prediction	Definition
Write key term here.	*Write what you think the term means in this column.*	*Define the term here. Add an example and use it in a sentence.*

Key Terms

Littoral zone	Mesotrophic	Coral reef
Limnetic zone	Eutrophic	Coral bleaching
Phytoplankton	Freshwater wetland	Open ocean
Profundal zone	Salt marsh	Photic zone
Benthic zone	Mangrove swamp	Aphotic zone
Oligotrophic	Intertidal zone	Chemosynthesis

WHILE YOU READ THE MODULE

Define Key Terms

When you come across a new key term while reading the module, copy the definition into the "Definition" column of your key terms table. Add an example and use the term in a sentence. Compare your initial ideas to the actual definition.

AFTER YOU READ THE MODULE

Review Key Terms

Match the key terms on the left with the definitions on the right.

_____1. Littoral zone

    a. The shallow zone of soil and water in lakes and ponds where most algae and emergent plants grow

_____2. Limnetic zone

    b. A marsh containing nonwoody emergent vegetation, found along the coast in temperate climates

_____3. Phytoplankton

    c. Describes a lake with a low level of productivity

_____4. Profundal zone

d. The most diverse marine biome on Earth, found in warm, shallow waters beyond the shoreline

_____5. Benthic zone

e. Deep ocean water, located away from the shoreline where sunlight can no longer reach the ocean bottom

_____6. Oligotrophic

f. A region of water where sunlight does not reach, below the limnetic zone in very deep lakes

_____7. Mesotrophic

g. Describes a lake with a high level of productivity

_____8. Eutrophic

h. A zone of open water in lakes and ponds

_____9. Freshwater wetlands

i. The deeper layer of ocean water that lacks sufficient sunlight for photosynthesis

_____10. Salt marsh

j. A phenomenon in which algae inside corals die, causing the corals to turn white

_____11. Mangrove swamp

k. Aquatic biome submerged or saturated part of each year, supports emergent vegetation

_____12. Intertidal zone

l. A process used by some bacteria in the ocean to generate energy with methane and hydrogen sulfide

_____13. Coral reef

m. Describes a lake with a moderate level of productivity

_____14. Coral bleaching

n. The upper layer of ocean water in the ocean that receives enough sunlight for photosynthesis

_____15. Open ocean

o. The narrow band of coastline between the levels of high tide and low tide

_____16. Photic zone

p. Floating algae

_____17. Aphotic zone

q. Swamp that occurs along tropical and subtropical coasts, contains salt-tolerant trees

_____18. Chemosynthesis

r. The muddy bottom of a lake, pond, or ocean

Chapter (4) Review Exercises

Check Your Understanding

Review "Learning Objectives Revisited" on page 141 of your textbook. Compare the notes you took while reading each module. Complete these exercises to review the chapter.

1. What are the five processes that determine climate?

2. Draw Figure 9.1 below making sure to label the altitudes on the y-axis and the temperatures on the x-axis.

3. Name the three phenomena that cause wind patterns worldwide.

4. Explain how Earth's oceans help regulate temperature on the planet.

5. List the major terrestrial biomes on Earth. Include a brief description of the general climate and plant growth for each biome.

6. What are the major aquatic biomes on Earth?

Practice for Free-Response Questions

Complete this exercise to build and practice the skills you will need to answer free-response questions on the exam. Use a separate sheet of paper if necessary.

> Biomes are characterized by distinct temperature and precipitation patterns. Explain how atmospheric convection currents determine the distribution of global biomes.

Review and Reflect

Complete these activities to solidify your knowledge of the chapter concepts and key terms. Use a notebook or a separate sheet of paper if necessary.

1. Review your key terms table for each module.

   (a) Which words did you already know? Which were new to you?
   (b) Write a new sentence using each key term.
   (c) Create a set of flash cards that includes each key term. Use the cards to review terms that were new or challenging.
   (d) When you feel comfortable with the new or challenging terms, review all of the cards, including those with familiar terms.
   (e) Save your cards to review before an exam.

2. What are the challenging concepts from this chapter?

   (a) Identify any concepts you found particularly challenging in this chapter.
   (b) Create a list of topics you need to review in preparation for an exam.

3. What questions do you have about concepts in the chapter?

   (a) Note any further questions you might have about material in the chapter.
   (b) Work with a partner to discuss these questions and ask your teacher for help as needed.

4. Write five possible multiple-choice questions based on this chapter. Work with a partner to quiz each other in preparation for an exam.

Chapter (5) Evolution of Biodiversity

Chapter Summary

This chapter describes the biodiversity of Earth, how it came to be, and how environmental factors can cause it to decline. Biodiversity can be described at the genetic level, the species level, and the ecosystem level. In any given location, biodiversity can be quantified by evaluating species richness and species evenness. Evolution is the underlying mechanism of biodiversity. Populations change over time in response to natural and artificial selection as well as random processes. Changes in the genetic composition of a population allow species to adapt to changing environmental conditions. Speciation occurs through allotropic and sympatric processes. Species evolve to exist in a unique niche that determines their geographic distribution. Environmental change can alter species distribution and lead to extinctions. The chapter consists of 4 modules:

- **Module 14**: The Biodiversity of Earth
- **Module 15**: How Evolution Creates Biodiversity
- **Module 16**: Speciation and the Pace of Evolution
- **Module 17**: Evolution of Niches and Species Distributions

Chapter Opening Case: *The Dung of the Devil*

The chapter opening case illustrates one of the most essential reasons to protect biodiversity: the use of resources for the development of new pharmaceutical drugs. Humans have already extracted life-saving drugs from a variety of species. With increasing rates of deforestation and habitat loss, many species that have never been researched for medical use are at risk of extinction. Furthermore, indigenous peoples with knowledge about medical uses of natural drugs are being forced to relocate. Their knowledge may soon be lost. This introductory case helps students understand the significance of biodiversity and the underlying mechanisms that allow organisms to adapt to their ever-changing environments.

Do the Math

This chapter contains the following "Do the Math" box to help prepare you for calculation questions you might encounter on the exam.

- "Measuring Species Diversity" (page 152)

To make sure you understand the concepts and techniques presented in these boxes, do the practice problems presented in the text as well as the additional "Practice the Math" problems that appear in Module 14 of this study guide.

Module 14: The Biodiversity of Earth

BEFORE YOU READ THE MODULE

Focus on Learning Objectives

Use the module learning objectives to guide your reading. On a separate piece of paper, write down each objective and take notes to help you meet each learning objective. After studying this module, you should be able to:

- understand how we estimate the number of species living on Earth.
- quantify biodiversity.
- describe patterns of relatedness among species using a phylogeny.

Preview Key Terms

In a notebook or on a separate sheet of paper, create a table like the one shown here to help with learning new key terms in the module. Before you read, fill out the "Prediction" column. Write what you think the term might mean or what it makes you think about. Use examples from your everyday life. There are no wrong answers!

Key Term	Prediction	Definition
Write key term here.	*Write what you think the term means in this column.*	*Define the term here. Add an example and use it in a sentence.*

Key Terms

Species richness
Species evenness
Phylogeny

WHILE YOU READ THE MODULE

Define Key Terms

When you come across a new key term while reading the module, copy the definition into the "Definition" column of your key terms table. Add an example and use the term in a sentence. Compare your initial ideas to the actual definition.

Practice the Math: Measuring Species Diversity

Read "Do the Math: Measuring Species Diversity." Try "Your Turn." For more math practice, do the following problems. Remember to show your work.

Using Shannon's index compare the species richness of the two communities described below.

1. Community A: 5 different species are evenly distributed among a community of 100 individuals.

2. Community B: 6 species are found in a community of 100. Four species are represented by 5 individuals. The remaining two species are evenly divided among the remaining population.

AFTER YOU READ THE MODULE

Review Key Terms

Match the key terms on the left with the definitions on the right.

_____1. Species richness

a. The relative proportion of individuals within the different species in a given area

_____2. Species evenness

b. The branching pattern of evolutionary relationships

_____3. Phylogeny

c. The number of species in a given area

Module 15: Environmental Indicators and Sustainability

BEFORE YOU READ THE MODULE

Focus on Learning Objectives

Use the module learning objectives to guide your reading. On a separate piece of paper, write down each objective and take notes to help you meet each learning objective. After studying this module, you should be able to:

- identify key environmental indicators and their trends over time.
- define sustainability and explain how it can be measured using the ecological footprint.

Preview Key Terms

In a notebook or on a separate sheet of paper, create a table like the one shown here to help with learning new key terms in the module. Before you read, fill out the "Prediction" column. Write what you think the term might mean or what it makes you think about. Use examples from your everyday life. There are no wrong answers!

Key Term	Prediction	Definition
Write key term here.	*Write what you think the term means in this column.*	*Define the term here. Add an example and use it in a sentence.*

Key Terms

Evolution

Microevolution

Macroevolution

Genes

Genotype

Phenotype

Mutation

Recombination

Evolution by artificial selection

Evolution by natural selection

Fitness

Adaptations

Gene flow

Genetic drift

Bottleneck effect

Founder effect

WHILE YOU READ THE MODULE

Define Key Terms

When you come across a new key term while reading the module, copy the definition into the "Definition" column of your key terms table. Add an example and use the term in a sentence. Compare your initial ideas to the actual definition.

Study the Figure

Evaluate the phylogenic tree in Figure 15.3, "Artificial selection on animals" on page 157.

1. Which of the breeds listed would be expected to have the greatest difference in genetic composition from their ancestral wolf? Which breeds would be most similar? Explain your reasoning.

AFTER YOU READ THE MODULE

Review Key Terms

Match the key terms on the left with the definitions on the right.

_____1. Evolution

_____2. Microevolution

_____3. Macroevolution

_____4. Gene

_____5. Genotype

_____6. Phenotype

_____7. Mutation

_____8. Recombination

_____9. Evolution by artificial selection

_____10. Evolution by natural selection

_____11. Fitness

a. The complete set of genes in an individual

b. The process in which humans determine which individuals breed, typically with a preconceived set of traits in mind

c. The genetic process by which one chromosome breaks off and attaches to another chromosome during reproductive cell division

d. A set of traits expressed by an individual

e. A change in the genetic composition of a population as a result of descending from a small number of colonizing individuals

f. An individual's ability to survive and reproduce

g. Evolution below the species level

h. The process by which individuals move from one population to another and thereby alter the genetic composition of both populations

i. A reduction in the genetic diversity of a population caused by a reduction in its size

j. A random change in the genetic code produced by a mistake in the copying process

k. The death of the last member of a species

_____12. Adaptation

l. A change in the genetic composition of a population over time

_____13. Gene flow

m. A change in the genetic composition of a population over time as a result of random mating

_____14. Genetic drift

n. Evolution that gives rise to new species, genera, families, classes, or phyla

_____15. Bottleneck effect

o. The process in which the environment determines which individuals survive and reproduce

_____16. Extinction

p. A physical location on the chromosomes within each cell of an organism

_____17. Founder effect

q. A trait that improves an individual's fitness

Module 16: Speciation and the Pace of Evolution

BEFORE YOU READ THE MODULE

Focus on Learning Objectives

Use the module learning objectives to guide your reading. On a separate piece of paper, write down each objective and take notes to help you meet each learning objective. After studying this module, you should be able to:

- explain the processes of allopatric and sympatric speciation.
- understand the factors that affect the pace of evolution.

Preview Key Terms

In a notebook or on a separate sheet of paper, create a table like the one shown here to help with learning new key terms in the module. Before you read, fill out the "Prediction" column. Write what you think the term might mean or what it makes you think about. Use examples from your everyday life. There are no wrong answers!

Key Term	Prediction	Definition
Write key term here.	_Write what you think the term means in this column._	_Define the term here. Add an example and use it in a sentence._

Key Terms

Geographic isolation
Reproductive isolation
Allopatric speciation

Sympatric speciation
Genetically modified organism

Define Key Terms

When you come across a new key term while reading the module, copy the definition into the "Definition" column of your key terms table. Add an example and use the term in a sentence. Compare your initial ideas to the actual definition.

AFTER YOU READ THE MODULE

Review Key Terms

Match the key terms on the left with the definitions on the right.

_____1. Geographic isolation

a. The result of two populations within a species evolving separately to the point that they can no longer interbreed and produce viable offspring

_____2. Allopatric speciation

b. Physical separation of a group of individuals from others of the same species

_____3. Reproductive isolation

c. The evolution of one species into two, without geographic isolation

_____4. Sympatric speciation

d. The process of speciation that occurs with geographic isolation

_____5. Genetically modified organism (GMO)

e. An organism produced by copying genes from a species with a desirable trait and inserting them into another species

Module 17: Evolution of Niches and Species Distributions

BEFORE YOU READ THE MODULE

Focus on Learning Objectives

Use the module learning objectives to guide your reading. On a separate piece of paper, write down each objective and take notes to help you meet each learning objective. After studying this module, you should be able to:

- explain the difference between a fundamental and a realized niche.
- describe how environmental change can alter species distributions.
- discuss how environmental change can cause species extinctions.

Preview Key Terms

In a notebook or on a separate sheet of paper, create a table like the one shown here to help with learning new key terms in the module. Before you read, fill out the "Prediction" column. Write what you think the term might mean or what it makes you think about. Use examples from your everyday life. There are no wrong answers!

Key Term	Prediction	Definition
Write key term here.	*Write what you think the term means in this column.*	*Define the term here. Add an example and use it in a sentence.*

Key Terms

Range of tolerance
Fundamental niche
Realized niche

Distribution
Niche generalists
Niche specialists

Mass extinction

WHILE YOU READ THE MODULE

Define Key Terms

When you come across a new key term while reading the module, copy the definition into the "Definition" column of your key terms table. Add an example and use the term in a sentence. Compare your initial ideas to the actual definition.

Study the Figure

Examine Figure 17.1, "Range of tolerance" on page 169 and answer the following questions. Use a separate piece of paper if necessary.

1. Describe the relationship between pH and the mortality of minnows.

2. If conditions exceed an organism's range of tolerance, how will population be affected?

3. Create a graph using the experimental data given below.

Measured pH of water	Number of minnows
2	0
4	230
6	561
8	598
10	199
12	0

4. Determine the minnow's optimal range of tolerance for pH.

Review Key Terms

Match the key terms on the left with the definitions on the right.

_____1. Range of tolerance

_____2. Fundamental niche

_____3. Realized niche

_____4. Distribution

_____5. Niche generalist

_____6. Niche specialist

_____7. Mass extinction

a. A species that can live under a wide range of abiotic or biotic conditions

b. Areas of the world in which a species lives

c. The suite of abiotic conditions under which a species can survive, grow, and reproduce

d. A species that is specialized to live in a specific habitat or to feed on a small group of species

e. A large extinction of species in a relatively short period of time

f. The limits to the abiotic conditions that a species can tolerate

g. The range of abiotic and biotic conditions under which a species actually lives

Chapter (5) Review Exercises

Check Your Understanding

Review "Learning Objectives Revisited" on page 27 of your textbook. Compare the notes you took while reading each module. Complete these exercises to review the chapter. Use a separate piece of paper if necessary.

1. Explain in your own words the difference between species evenness and species richness.

2. What are the key ideas of the theory of evolution by natural selection?

3. List and describe the five random processes that can drive evolution.

4. Summarize in your own words the example of allopatric speciation from Figure 16.1.

5. What factors determine the pace of evolution?

6. What five factors do scientists contend may be causing the sixth mass extinction?

Practice for Free-Response Questions

Complete this exercise to build and practice the skills you will need to answer free-response questions on the exam. Use a separate sheet of paper if necessary.

Evolution by natural selection is the process by which the environment determines which individuals survive and reproduce. Explain how the allele frequency of an amphipod population changes in response to the introduction of a predatory fish.

Review and Reflect

Complete these activities to solidify your knowledge of the chapter concepts and key terms. Use a notebook or a separate sheet of paper if necessary.

1. Review your key terms table for each module.

   (a) Which words did you already know? Which were new to you?
   (b) Write a new sentence using each key term.
   (c) Create a set of flash cards that includes each key term. Use the cards to review terms that were new or challenging.
   (d) When you feel comfortable with the new or challenging terms, review all of the cards, including those with familiar terms.
   (e) Save your cards to review before an exam.

2. What are the challenging concepts from this chapter?

   (a) Identify any concepts you found particularly challenging in this chapter.
   (b) Create a list of topics you need to review in preparation for an exam.

3. What questions do you have about concepts in the chapter?

   (a) Note any further questions you might have about material in the chapter.
   (b) Work with a partner to discuss these questions and ask your teacher for help as needed.

4. Write five possible multiple-choice questions based on this chapter. Work with a partner to quiz each other in preparation for an exam.

Unit 2 Multiple-Choice Review Exam

Choose the best answer.

1. Which is the correct equation for photosynthesis?
   (A) Energy + $6H_2O$ + 7 CO_2 → $C_6H_{12}O_6$ + $8O_2$
   (B) Energy + 6 H_2O + 6 CO_2 → $C_6H_{12}O_6$ + 6 O_2
   (C) Solar energy + 6 H_2O + 8 CO_2 → $C_6H_{12}O_6$ + 8 O_2
   (D) Solar energy + 8 H_2O + 8 CO_2 → $C_6H_{12}O_6$+ 12 O_2
   (E) Solar energy + 6 H_2O + 6 CO_2 → $C_6H_{12}O_6$ + 6 O_2

2. Which is the correct flow of energy in an ecosystem?
   (A) Producer → Herbivore → Carnivore → Scavenger
   (B) Producer → Scavenger → Herbivore → Carnivore
   (C) Scavenger → Carnivore → Herbivore → Producer
   (D) → Scavenger → Herbivore → Producer → Carnivore
   (E) Scavenger → Producer → Herbivore → Carnivore

3. The net primary productivity of an ecosystem is 25 kg C/m^2/year, and the energy needed by the producers for their own respiration is 30 kg C/m^2/year. The gross primary productivity of such an ecosystem would be
   (A) 5 kg C/ m^2/year.
   (B) 10 kg C/ m^2/year.
   (C) 25 kg C/ m^2/year.
   (D) 55 kg C/ m^2/year.
   (E) 30 kg C/ m^2/year.

4. An ecosystem has an ecological efficiency of 10 percent. If the tertiary consumer level has 1 kcal of energy, how much energy did the producer level contain?
   (A) 100 kcal
   (B) 1,000 kcal
   (C) 10,000 kcal
   (D) 90 kcal
   (E) 900 kcal

5. Which of the following is NOT a part of the hydrologic cycle?
   (A) Transpiration
   (B) Sedimentation
   (C) Evapotranspiration
   (D) Runoff
   (E) Infiltration

Match the following processes with the correct product.

6. _____ Nitrogen fixation

7. _____ Nitrification

8. _____ Assimilation

9. _____ Ammonification

10. _____ Mineralization

(A) Nitrogen is assimilated into plant tissues.

(B) Organic matter is converted back into inorganic compounds.

(C) Ammonia is converted into nitrite and then nitrate.

(D) Bacteria convert ammonia into ammonium.

(E) Decomposers use waste as a food source and excrete ammonium.

11. In which layer of Earth's atmosphere does the ozone layer appear?
    (A) Troposphere
    (B) Stratosphere
    (C) Mesosphere
    (D) Thermosphere
    (E) Exosphere

12. Climate is
    (A) the amount of annual precipitation, the average temperature, and humidity.
    (B) the albedo and the average precipitation.
    (C) the conditions created by adiabatic cooling and heating.
    (D) the saturation point of the air and the average temperature in a region.
    (E) the average weather that occurs in a given region over a long period of time.

13. The prevailing wind systems of the world are produced by
    (A) convection currents and the Coriolis effect.
    (B) convection currents and adiabatic cooling.
    (C) convection currents and adiabatic heating.
    (D) the Earth's rotation and the Coriolis effect.
    (E) ocean circulation and the Coriolis effect.

14. During which months is the Sun directly overhead at the equator?
    (A) January and June
    (B) March and September
    (C) December and June
    (D) July and February
    (E) All months

15. The El Niño-Southern Oscillation brings what type of weather conditions to the southeastern United States?
    (A) Warmer, drier
    (B) Warmer, wetter
    (C) Cooler, drier
    (D) Cooler, wetter
    (E) Warmer, windier

Use the following climate diagram to answer questions 16-17.

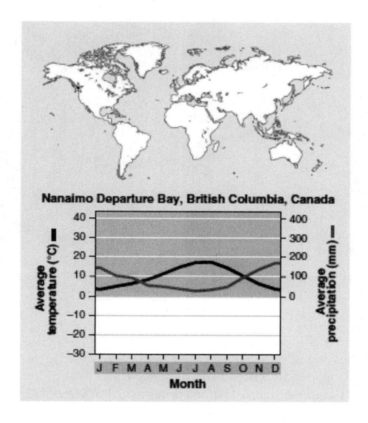

16. When is temperature at the highest and precipitation at the lowest?
    (A) January
    (B) April
    (C) July
    (D) October
    (E) December

17. In which month is average rainfall approximately 20 mm?
    (A) January
    (B) March
    (C) July
    (D) October
    (E) December

18. Which of the following is NOT used to determine biodiversity?
   (A) Species evenness
   (B) Shannon's index
   (C) Genetic composition
   (D) Species richness
   (E) Species range of tolerance

Use the figure to answer question 19.

Community 1
A: 25%   B: 25%   C: 25%   D: 25%

Community 2
A: 70%   B: 10%   C: 10%   D: 10%

Figure 14.2
Environmental Science for AP®, Second Edition
© 2015 W.H. Freeman and Company

19. Which of the following describes the picture above?
   (A) Community 1 is less diverse than community 2.
   (B) Community 1 is more diverse than community 2.
   (C) Community 1 has a greater evenness and a greater richness than community 2.
   (D) Community 1 has a lower evenness and a lower richness than community 2.
   (E) Community 2 has a greater evenness but a lower richness than community 1.

20. Terrestrial biomes are defined by which parameter?
   I.   Average temperature and precipitation
   II. Distinctive plant types
   III. Distinctive animal species

   (A) I only
   (B) II only
   (C) I and II
   (D) I and III
   (E) I, II and III

21. Speciation that occurs as a result of geographic isolation is called
    (A) gene flow.
    (B) allopatric speciation.
    (C) sympatric speciation.
    (D) bottleneck effect.
    (E) founder effect.

22. Which is a benefit of genetically modified plants?
    I. They are more resistant to pests.
    II. They offer higher yields.
    III. They allow species to find new niches.
    (A) I
    (B) I and II
    (C) I, II, and III
    (D) I and III
    (E) II and III

23. A fundamental niche is
    (A) the abiotic conditions under which a species can survive, grow, and reproduce.
    (B) the areas of the world in which a species lives.
    (C) the limits to the abiotic conditions a species can tolerate.
    (D) the range of abiotic and biotic conditions under which a species actually lives.
    (E) the place where a niche specialist can best survive.

24. In which type of rock would you most likely find a fossil?
    (A) Igneous
    (B) Sedimentary
    (C) Metamorphic
    (D) Mineral
    (E) Marble

25. Which factor is NOT likely to contribute significantly to a mass extinction?
    (A) Habitat destruction
    (B) Overharvesting
    (C) Invasive species
    (D) Genetic breeding
    (E) Climate change

26. This terrestrial biome is characterized by cold winters and short growing seasons. The dominant plant life includes cone-bearing trees. Temperature in this region limits plant growth. Which biome fits this description?
    (A) The tundra
    (B) The boreal forest
    (C) The temperate rainforest
    (D) The temperate seasonal forest
    (E) The cold desert

27. Which two terrestrial biomes are often used for agriculture because of high soil fertility?
    (A) Tropical rainforests and temperate grasslands
    (B) Tropical grasslands and temperate rainforests
    (C) Tropical rainforests and temperate rainforests
    (D) Temperate grasslands and temperate seasonal forests
    (E) Boreal forest and temperate seasonal forests

28. Which describes ecosystem services associated with freshwater wetlands?
    I.   Wetlands absorb excess rainfall and reduce flooding.
    II. Wetlands filter pollutants from water.
    III. Wetlands provide critical breeding habitats for migratory birds.
    (A) I only
    (B) II only
    (C) III only
    (D) I and III
    (E) I, II, and III

29. Due to a drought, a lake dries into two small, independent lakes. Over time natural selection favors a different set of traits for a single species of fish found in both lakes. After a flood, the lakes are reconnected and fish populations rejoin but do not breed. Which process has occurred?
    (A) Genetic drift
    (B) The bottleneck effect
    (C) Sympatric speciation
    (D) Recombination
    (E) Allopatric speciation

30. Which is NOT true regarding the process of sympatric speciation?
    (A) Sympatric speciation is a process that creates new species through geographic isolation.
    (B) Sympatric speciation results in polyploidy.
    (C) Sympatric speciation results in two species that cannot breed with each other.
    (D) Sympatric speciation has been encouraged by plant breeders to produce larger fruit.
    (E) Sympatric speciation results in reproductive isolation.

Chapter (6) Population and Community Ecology

Chapter Summary

This chapter examines population characteristics, factors that drive populations to change, and different reproductive strategies. It looks at community relationships, how members of a community share resources, and how ecosystems evolve over time. The chapter presents both exponential growth and logistic growth and their relationship to carrying capacity. Limiting resources typically prevent exponential growth. Environmental pressures often cause population growth to stabilize over time to a sustainable level. We call this level carrying capacity. The logistic growth model, an S-shaped curve, depicts the relationship between population growth and carrying capacity. After studying this chapter you should be able to graph the growth models, survivorship curves, and predator–prey relationships. In addition, you should have a firm understanding of population distribution, growth models, r-selected and K-selected species, outcomes of competition, and primary versus secondary succession. The chapter consists of 4 modules:

- **Module 18:** The Abundance and Distribution of Populations
- **Module 19:** Population Growth Models
- **Module 20:** Community Ecology
- **Module 21:** Community Succession

Chapter Opening Case: *New England Forests Come Full Circle*

The chapter opening case illustrates the complexity of community interactions and provides an historical example of ecological recovery and secondary succession. The story follows the New England ecosystem through a series of changes including deforestation and conversion to croplands, field abandonment, and the reestablishment of grasses and trees. This story exemplifies the multifaceted interactions among species that drive the ever-changing ecosystem composition.

Do the Math

This chapter contains the following "Do the Math" box to help prepare you for calculation questions you might encounter on the exam.

- "Calculating Exponential Growth" (page 199)

To make sure you understand the concepts and techniques presented in these boxes, do the practice problems presented in the text as well as the additional "Practice the Math" problems that appear in Module 19 of this study guide.

Module 18: The Abundance and Distribution of Populations

BEFORE YOU READ THE MODULE

Focus on Learning Objectives

Use the module learning objectives to guide your reading. On a separate piece of paper, write down each objective and take notes to help you meet each learning objective. After studying this module, you should be able to:

- explain how nature exists at several levels of complexity.
- discuss the characteristics of populations.
- contrast the effects of density-dependent and density-independent factors on population growth.

Preview Key Terms

In a notebook or on a separate sheet of paper, create a table like the one shown here to help with learning new key terms in the module. Before you read, fill out the "Prediction" column. Write what you think the term might mean or what it makes you think about. Use examples from your everyday life. There are no wrong answers!

Key Term	Prediction	Definition
Write key term here.	*Write what you think the term means in this column.*	*Define the term here. Add an example and use it in a sentence.*

Key Terms

Population

Community

Population ecology

Population size (*N*)

Population density

Population distribution

Sex ratio

Age structure

Limiting resource

Density-dependent factor

Carrying capacity (*K*)

Density-independent factor

WHILE YOU READ THE MODULE

Define Key Terms

When you come across a new key term while reading the module, copy the definition into the "Definition" column of your key terms table. Add an example and use the term in a sentence. Compare your initial ideas to the actual definition.

AFTER YOU READ THE MODULE

Review Key Terms

Match the key terms on the left with the definitions on the right.

_____1. Population

_____2. Community

_____3. Population ecology

_____4. Population size (*N*)

_____5. Population density

_____6. Population distribution

_____7. Sex ratio

_____8. Age structure

_____9. Limiting resource

_____10. Density-dependent factor

_____11. Carrying capacity *(K)*

_____12. Density-independent factor

a. The ratio of males to females in a population

b. A description of how many individuals fit into particular age categories in a population

c. The number of individuals per unit area at a given time

d. A factor that influences an individual's probability of survival and reproduction in a manner that depends on the size of the population

e. The individuals that belong to the same species and live in a given area at a particular time

f. The total number of individuals within a defined area at a given time

g. All of the populations of organisms within a given area

h. A description of how individuals are distributed with respect to one another

i. A resource that a population cannot live without and that occurs in quantities lower than the population would require to increase in size

j. A factor that has the same effect on an individual's probability of survival and the amount of reproduction at any population size

k. The limit of how many individuals in a population the environment can sustain

l. The study of factors that cause populations to increase or decrease

Module 19: Population Growth Models

BEFORE YOU READ THE MODULE

Focus on Learning Objectives

Use the module learning objectives to guide your reading. On a separate piece of paper, write down each objective and take notes to help you meet each learning objective. After studying this module, you should be able to:

- explain the exponential growth model of populations, which produces a J-shaped curve.
- describe how the logistic growth model incorporates a carrying capacity and produces an S-shaped curve.
- compare the reproductive strategies and survivorship curves of different species.
- explain the dynamics that occur in metapopulations.

Preview Key Terms

In a notebook or on a separate sheet of paper, create a table like the one shown here to help with learning new key terms in the module. Before you read, fill out the "Prediction" column. Write what you think the term might mean or what it makes you think about. Use examples from your everyday life. There are no wrong answers!

Key Term	Prediction	Definition
Write key term here.	Write what you think the term means in this column.	Define the term here. Add an example and use it in a sentence.

Key Terms

Density-independent factor
Population growth model
Population growth rate
Intrinsic growth rate (r)
Exponential growth model ($N_t = N_0 e^{rt}$)
J-shaped curve
Logistic growth model
S-shaped curve
Overshoot
Die-off

K-selected species
r-selected species
Survivorship curve
Type I survivorship curve
Type II survivorship curve
Type III survivorship curve
Corridor
Metapopulation
Inbreeding depression

Define Key Terms

When you come across a new key term while reading the module, copy the definition into the "Definition" column of your key terms table. Add an example and use the term in a sentence. Compare your initial ideas to the actual definition.

Study the Figure

Examine Figure 19.2, "The logistic growth model" on page 198. Carrying capacity is determined by numerous density-dependent factors.

1. Explain how an S-shaped curve depicts a pattern of population growth where resources are not unlimited.

Practice the Math: Calculating Exponential Growth

Read "Do the Math: Calculating Exponential Growth," on page 199. Try "Your Turn." For more math practice, do the following problems. Remember to show your work. (Note that you will need to use a calculator to complete these problems.)

1. A population of deer has an initial population size of 15 individuals (N_0=15). Assume that the intrinsic rate of growth for a deer is $r = 0.25$ or 25%, which means that each deer produces a net increase of 0.25 deer each year. (Formula: $N_t = N_0 e^{rt}$)

   With this information, predict the size of the deer population in 1 year, 5 years, and 10 years.

2. Assume that the intrinsic rate of growth is 0.75 for mosquitoes. Calculate the predicted size of the mosquito population after 1, 5, and 10 years if the initial population size is 100 individuals.

Review Key Terms

Match the key terms on the left with the definitions on the right.

_____1. Population growth models

a. The shape of the logistic growth model when graphed

_____2. Population growth rate

b. The curve of the exponential growth model when graphed

_____3. Intrinsic growth rate

c. A species with a low intrinsic growth rate that causes the population to increase slowly until it reaches carrying capacity

_____4. Exponential growth model ($N_t = N_0e^{rt}$)

d. A rapid decline in a population due to death

_____5. J-shaped curve

e. A graph that represents the distinct patterns of species survival as a function of age

_____6. Logistic growth model

f. The number of offspring an individual can produce in a given time period, minus the deaths of the individual or its offspring during the same period

_____7. S-shaped curve

g. Strips of natural habitat that connect populations

_____8. Overshoot

h. A pattern of survival over time in which there is low survivorship early in life with few individuals reaching adulthood

_____9. Die-off

i. When individuals with similar genotypes—typically relatives—breed with each other and produce offspring that have an impaired ability to survive and reproduce

_____10. *K*-selected species

j. A growth model that estimates a population's future size (N_t) after a period of time (t), based on the intrinsic growth rate (r) and the number of reproducing individuals currently in the population (N_0)

_____11. *r*-selected species

k. A growth model that describes a population whose growth is initially exponential, but slows as the population approaches the carrying capacity of the environment

_____12. Survivorship curve

l. A group of spatially distinct populations that are connected by occasional movements of individuals between them

_____13. Type I survivorship curve

m. Mathematical equations that can be used to predict population size at any moment in time

_____14. Type II survivorship curve

n. A pattern of survival over time in which there is high survival throughout most of the life span, but then individuals start to die in large numbers as they approach old age

_____15. Type III survivorship curve

o. When a population becomes larger than the environment's carrying capacity

_____16. Corridor

p. A pattern of survival over time in which there is a relatively constant decline in survivorship throughout most of the life span

_____17. Metapopulation

q. The maximum potential for growth of a population under ideal conditions with unlimited resources

_____18. Inbreeding depression

r. A species that has a high intrinsic growth rate, which often leads to population overshoots and die-offs

Module 20: Community Ecology

BEFORE YOU READ THE MODULE

Focus on Learning Objectives

Use the module learning objectives to guide your reading. On a separate piece of paper, write down each objective and take notes to help you meet each learning objective. After studying this module, you should be able to:

- identify species interactions that cause negative effects on one or both species.
- discuss species interactions that cause neutral or positive effects on both species.
- explain the role of keystone species.

Preview Key Terms

In a notebook or on a separate sheet of paper, create a table like the one shown here to help with learning new key terms in the module. Before you read, fill out the "Prediction" column. Write what you think the term might mean or what it makes you think about. Use examples from your everyday life. There are no wrong answers!

Key Term	Prediction	Definition
Write key term here.	*Write what you think the term means in this column.*	*Define the term here. Add an example and use it in a sentence.*

Key Terms

Community ecology
Symbiotic relationship
Competition
Competitive exclusion principle
Resource partitioning
Predation
Parasitoid

Parasitism
Pathogen
Herbivory
Mutualism
Commensalism
Keystone species
Ecosystem engineer

WHILE YOU READ THE MODULE

Define Key Terms

When you come across a new key term while reading the module, copy the definition into the "Definition" column of your key terms table. Add an example and use the term in a sentence. Compare your initial ideas to the actual definition.

Review Key Terms

Match the key terms on the left with the definitions on the right.

_____1. Community ecology

_____2. Symbiotic relationship

_____3. Competition

_____4. Competitive exclusion principle

_____5. Resource partitioning

_____6. Predation

_____7. Parasitoid

_____8. Parasitism

_____9. Pathogen

_____10. Herbivory

_____11. Mutualism

_____12. Commensalism

_____13. Keystone species

_____14. Ecosystem engineer

a. The study of interactions between species

b. A specialized type of predator that lays eggs inside other organisms—referred to as its host

c. An interaction in which an animal consumes a producer

d. The struggle of individuals to obtain a shared limiting resource

e. A species that plays a far more important in its community than its relative abundance might suggest

f. A parasite that causes disease in its host

g. The principle stating that two species competing for the same limiting resource cannot coexist

h. A keystone species that creates or maintains habitat for other species

i. An interaction between two species that increases the chances of survival or reproduction for both species

j. An interaction in which one animal typically kills and consumes another animal

k. A relationship between species in which one species benefits and the other species is neither harmed nor helped

l. The relationship between two species that live in close association with each other

m. An interaction in which one organism lives on or in another organism

n. When two species divide a resource based on differences in their behavior or morphology

Module 21: Community Succession

BEFORE YOU READ THE MODULE

Focus on Learning Objectives

Use the module learning objectives to guide your reading. On a separate piece of paper, write down each objective and take notes to help you meet each learning objective. After studying this module, you should be able to:

- explain the process of primary succession.
- explain the process of secondary succession.
- explain the process of aquatic succession.
- describe the factors that determine the species richness of a community.

Preview Key Terms

In a notebook or on a separate sheet of paper, create a table like the one shown here to help with learning new key terms in the module. Before you read, fill out the "Prediction" column. Write what you think the term might mean or what it makes you think about. Use examples from your everyday life. There are no wrong answers!

Key Term	Prediction	Definition
Write key term here.	*Write what you think the term means in this column.*	*Define the term here. Add an example and use it in a sentence.*

Key Terms

Ecological succession
Primary succession
Secondary succession

Pioneer species
Theory of island biogeography

WHILE YOU READ THE MODULE

Define Key Terms

When you come across a new key term while reading the module, copy the definition into the "Definition" column of your key terms table. Add an example and use the term in a sentence. Compare your initial ideas to the actual definition.

Study the Figure

Examine Figure 21.4, "Habitat size and species richness" on page 216.

1. State the relationship between island size and the number of bird species. According to the theory of island biogeography, which other factor contributes to greater species diversity?

Chapter (6) Review Exercises

Check Your Understanding

Review "Learning Objectives Revisited" on page 27 of your textbook. Compare the notes you took while reading each module. Complete these exercises to review the chapter.

Check Your Understanding

1. Diagram the different population distributions in the boxes below.

      Random   **Uniform**   **Clumped**

2. Using examples, describe and compare density-dependent factors and density-independent factors.

3. Draw a J-shaped growth curve and a logistic growth curve in the boxes below.

J-shaped

Logistic

4. Draw a typical carrying capacity graph where the population grows exponentially, overshoots the carrying capacity, has a die-off and then hovers around the carrying capacity.

5. Describe the predator-prey relationship in Figure 19.5.

6. Compare mutualism, commensalism, and parasitism by completing the table below.

Relationship	Definition	Example
Mutualism		
Commensalism		
Parasitism		

7. Explain the difference between primary succession and secondary succession by considering the cause of each and the types of plant species typical of each.

Practice for Free-Response Questions

Complete this exercise to build and practice the skills you will need to answer free-response questions on the exam. Use a separate sheet of paper if necessary.

1. Draw a graph that represents population growth limited by density dependent factors. Explain how reproductive strategies of a *K*-selected species are reflected in this growth model

2. Draw a second graph that represents a population growing with unlimited resources but then experiences a die-off event due to density independent factors. Explain how reproductive strategies of an *r*-selected species are reflected in this growth model.

Review and Reflect

Complete these activities to solidify your knowledge of the chapter concepts and key terms. Use a notebook or a separate sheet of paper if necessary.

1. Review your key terms table for each module.

   (a) Which words did you already know? Which were new to you?
   (b) Write a new sentence using each key term.
   (c) Create a set of flash cards that includes each key term. Use the cards to review terms that were new or challenging.
   (d) When you feel comfortable with the new or challenging terms, review all of the cards, including those with familiar terms.
   (e) Save your cards to review before an exam.

2. What are the challenging concepts from this chapter?

   (a) Identify any concepts you found particularly challenging in this chapter.
   (b) Create a list of topics you need to review in preparation for an exam.

3. What questions do you have about concepts in the chapter?

   (a) Note any further questions you might have about material in the chapter.
   (b) Work with a partner to discuss these questions and ask your teacher for help as needed.

4. Write five possible multiple-choice questions based on this chapter. Work with a partner to quiz each other in preparation for an exam.

Chapter (7) The Human Population

Chapter Summary

This chapter covers human population growth, including the social, economic, and environmental factors that determine growth. This chapter also addresses the relationship between resource consumption and population growth. After studying this chapter, you should know how to calculate population growth rates and interpret age structure diagrams. You should understand the phases of the demographic transition model and be able to correlate economic development to family size. Other essential topics in this chapter are the relationship between education and total fertility rates and the environmental impacts of both poverty and affluence. The chapter consists of 2 modules:

- **Module 22:** Human Population Numbers
- **Module 23:** Economic Development, Consumption, and Sustainability

Chapter Opening Case: *The Environmental Implications of China's Growing Population*

The chapter opening case provides an introductory discussion about the topics addressed in Chapter 7. It offers will a first glimpse at China's large population numbers and the country's growing affluence. The one-child policy is introduced in this case. The case also discusses how China's growing affluence affects resource use and pollution. This case provides an opportunity to introduce differences in resource use between developed and developing nations as well as environmental impacts as consumption increases across the world.

Do the Math

This chapter contains the following "Do the Math" box to help prepare you for calculation questions you might encounter on the exam.

- "Calculating Population Growth" (page 233)

To make sure you understand the concepts and techniques presented in these boxes, do the practice problems presented in the text as well as the additional "Practice the Math" problems that appear in Module 22 of this study guide.

Module 22: Human Population Numbers

BEFORE YOU READ THE MODULE

Focus on Learning Objectives

Use the module learning objectives to guide your reading. On a separate piece of paper, write down each objective and take notes to help you meet each learning objective. After studying this module, you should be able to:

- explain factors that may potentially limit the carrying capacity of humans on Earth.
- describe the drivers of human population growth.
- read and interpret an age structure diagram.

Preview Key Terms

In a notebook or on a separate sheet of paper, create a table like the one shown here to help with learning new key terms in the module. Before you read, fill out the "Prediction" column. Write what you think the term might mean or what it makes you think about. Use examples from your everyday life. There are no wrong answers!

Key Term	Prediction	Definition
Write key term here.	Write what you think the term means in this column.	Define the term here. Add an example and use it in a sentence.

Key Terms

Demography
Demographer
Immigration
Emigration
Crude birth rate (CBR)
Crude death rate (CDR)

Doubling time
Total fertility rate (TFR)
Replacement-level fertility
Developed country
Developing country
Life expectancy

Infant mortality
Child mortality
Net migration rate
Age structure diagram
Population pyramid
Population momentum

WHILE YOU READ THE MODULE

Define Key Terms

When you come across a new key term while reading the module, copy the definition into the "Definition" column of your key terms table. Add an example and use the term in a sentence. Compare your initial ideas to the actual definition.

Practice the Math: Calculating Population

Read "Do the Math: Calculating Population Growth," on page 233. Try "Your Turn." For more math practice, do the following problem. Remember to show your work. Use a separate piece of paper if necessary.

> In 2014, Lebanon had a population of 5.9 million people. The crude birth rate was 15, while the crude death rate was 5. The net migration rate was 84 per 1000. How many people will be added to the population if this trend continues over the next year? If these statistics remain the same, when will the population of Lebanon double?

Study the Figure

Examine Figure 22.8, "Age structure diagrams " on page 235.

1. Describe the age structure diagram for each country. What characteristics account for the trends in age structure that we see in each diagram?

AFTER YOU READ THE MODULE

Review Key Terms

Match the key terms on the left with the definitions on the right.

_____1. Demography   a. The number of years it takes a population to double

_____2. Demographer   b. A country with relatively high levels of industrialization and income

_____3. Immigration   c. A visual representation of the number of individuals within specific age groups for a country, typically expressed for males and females

_____4. Emigration

_____5. Crude birth rate (CBR)

_____6. Crude death rate (CDR)

_____7. Doubling time

_____8. Total fertility rate (TFR)

_____9. Replacement-level fertility

_____10. Developed country

_____11. Developing country

_____12. Life expectancy

_____13. Infant mortality

_____14. Child mortality

_____15. Net migration rate

_____16. Age structure diagram

_____17. Population pyramid

_____18. Population momentum

d. The number of deaths of children under age 5 per 1,000 live births

e. An age structure diagram that is widest at the bottom and smallest at the top, typical of developing countries

f. A scientist in the field of demography

g. A country with relatively low levels of industrialization and income

h. The movement of people out of a country or region

i. An estimate of the average number of children that each woman in a population will bear throughout her childbearing years

j. The number of births per 1,000 individuals per year

k. The number of deaths per 1,000 individuals per year

l. Continued population growth after growth reduction measures have been implemented

m. The movement of people into a country or region, from another country or region

n. The average number of years that an infant born in a particular year in a particular country can be expected to live, given the current average life span and death rate in that country

o. The study of human populations and population trends

p. The total fertility rate required to offset the average number of deaths in a population in order to maintain the current population size

q. The difference between immigration and emigration in a given year per 1,000 people in a country

r. The number of deaths of children under 1 year of age per 1,000 live births

Module 23: Economic Development, Consumption, and Sustainability

BEFORE YOU READ THE MODULE

Focus on Learning Objectives

Use the module learning objectives to guide your reading. On a separate piece of paper, write down each objective and take notes to help you meet each learning objective. After studying this module, you should be able to:

- describe how demographic transition follows economic development.
- explain how relationships among population size, economic development, and resource consumption influence the environment.
- describe why sustainable development is a common but elusive goal.

Preview Key Terms

In a notebook or on a separate sheet of paper, create a table like the one shown here to help with learning new key terms in the module. Before you read, fill out the "Prediction" column. Write what you think the term might mean or what it makes you think about. Use examples from your everyday life. There are no wrong answers!

Key Term	Prediction	Definition
Write key term here.	*Write what you think the term means in this column.*	*Define the term here. Add an example and use it in a sentence.*

Key Terms

Theory of demographic transition

Family planning

Affluence

IPAT equation

Gross domestic product (GDP)

Urban area

WHILE YOU READ THE MODULE

Define Key Terms

When you come across a new key term while reading the module, copy the definition into the "Definition" column of your key terms table. Add an example and use the term in a sentence. Compare your initial ideas to the actual definition.

Study The Figure

Examine Figure 23.4, "Total fertility rates for educated and uneducated women in six countries" on page 240.

1. Define total fertility rate. What is the relationship between TFR and education for women? Propose a reason for this correlation.

AFTER YOU READ THE MODULE

Review Key Terms

Match the key terms on the left with the definitions on the right.

_____1. Theory of demographic transition

a. measure of the value of all products and services produced in one year in one country

_____2. Affluence

b. The theory that as a country moves from a subsistence economy to industrialization and increased affluence it undergoes a predictable shift in population growth

_____3. Family planning

c. The practice of regulating the number or spacing of offspring through the use of birth control

_____4. IPAT equation

d. The state of having plentiful wealth including the possession of money, goods, or property

_____5. Gross domestic product (GDP)

e. An area that contains more than 385 people per square kilometer (1,000 people per square mile)

_____6. Urban area

f. An equation used to estimate the impact of the human lifestyle on the environment: Impact = population × affluence × technology

Chapter (7) Review Exercises

Check Your Understanding

Review "Learning Objectives Revisited" on page 249 of your textbook. Compare the notes you took while reading each module. Complete these exercises to review the chapter. Use a separate sheet of paper if needed.

1. Using the growth rate formula calculate the growth rate of a country with a CBR of 15, a CDR of 10, an immigration rate of 5, and an emigration rate of 2.

    Formula:
    $$\text{growth rate} = \frac{(\text{crude birth rate} + \text{immigration}) - (\text{crude death rate} + \text{emigration})}{10}$$

2. Calculate the doubling time of the country in question 1 above.

    Formula: $\text{Doubling time (in years)} = \dfrac{70}{\text{Growth rate}}$

Practice for Free-Response Questions

Complete this exercise to build and practice the skills you will need to answer free-response questions on the exam. Use a separate sheet of paper if needed.

Age structure diagrams quickly communicate demographics and trends within a population. Complete the table below to identify characteristics of a population that are associated with each type of age structure diagram.

Diagram	Population Characteristics
Population pyramid	
Even age distribution	
Inverted pyramid	

Review and Reflect

Complete these activities to solidify your knowledge of the chapter concepts and key terms. Use a notebook or a separate sheet of paper if necessary.

1. Review your key terms table for each module.

   (a) Which words did you already know? Which were new to you?
   (b) Write a new sentence using each key term.
   (c) Create a set of flash cards that includes each key term. Use the cards to review terms that were new or challenging.
   (d) When you feel comfortable with the new or challenging terms, review all of the cards, including those with familiar terms.
   (e) Save your cards to review before an exam.

2. What are the challenging concepts from this chapter?

   (a) Identify any concepts you found particularly challenging in this chapter.
   (b) Create a list of topics you need to review in preparation for an exam.

3. What questions do you have about concepts in the chapter?

   (a) Note any further questions you might have about material in the chapter.
   (b) Work with a partner to discuss these questions and ask your teacher for help as needed.

4. Write five possible multiple-choice questions based on this chapter. Work with a partner to quiz each other in preparation for an exam.

Unit 3 Multiple-Choice Review Exam

Choose the best answer.

1. Which is NOT a characteristic of a population?
   (A) The size of a population can change in response to environmental factors.
   (B) A population can remain relatively stable over time.
   (C) Population distributions may be random, uniform, or clumped.
   (D) A population consists of all the members of a particular species located on Earth.
   (E) Populations contain individuals of varying ages.

2. If a population of 300 deer increases to 400 deer, the percent change is
   (A) 3 percent.
   (B) 33 percent.
   (C) 300 percent.
   (D) 50 percent.
   (E) 25 percent.

3. Which is an *r*-selected species?
   (A) Humans
   (B) Elephants
   (C) Cockroaches
   (D) Dogs
   (E) Oak trees

4. Which refers to the maximum growth potential of a population under ideal conditions with unlimited resources?
   (A) Intrinsic growth rate
   (B) Exponential growth
   (C) Carrying capacity
   (D) Logistic growth
   (E) Population oscillation

5. Predation occurs when
   (A) one organism lives in or on another organism.
   (B) one animal kills and consumes another animal.
   (C) an animal kills and consumes a producer.
   (D) two species have an interaction that benefits their chances of survival.
   (E) one species benefits another and the second species is neither harmed nor helped.

6. When a beaver creates new habitat in an area, it is functioning as
    (A) a parasite.
    (B) an indicator species.
    (C) a mutualist.
    (D) an herbivore.
    (E) a keystone species.

7. During primary succession
    (A) plants colonize soil that remains in areas that have been disturbed.
    (B) grasses and wildflowers colonize new areas.
    (C) early-arriving plants colonize bare rock.
    (D) pioneer species dominate because of their ability to grow well in full sunshine.
    (E) wind-pollinated species predominate.

8. According to the theory of Island biogeography, which factors determine species richness?
    I. Island size
    II. Distance from the mainland
    III. Succession rates

    (A) I and II
    (B) II and III
    (C) I and III
    (D) I only
    (E) I, II, and III

9. Which is an example of resource partitioning?
    (A) Wolves and foxes prey on rabbits.
    (B) Squirrels hide acorns and seeds for the winter.
    (C) Blue jays prevent smaller birds from eating at a bird feeder.
    (D) An orchid lives on a tree.
    (E) Wolves hunt at dawn and dusk and coyotes hunt at night.

Use the following figure for question 10.

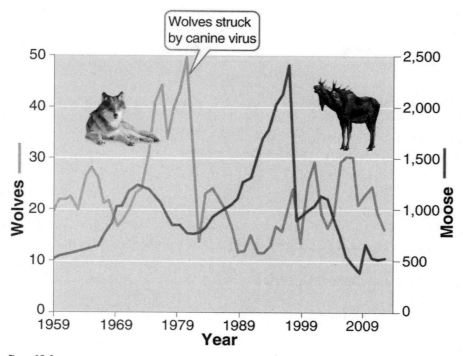

Figure 19.6
Environmental Science for AP®, Second Edition
Data from J. A. Vucetich and R. O. Peterson, Ecological Studies of Wolves on Isle Royale: Annual Report 2007–2008, School of Forest Resources and Environmental Science, Michigan Technological University

10. The graph illustrates
    (A) a mutualistic relationship.
    (B) exponential growth.
    (C) logistic growth.
    (D) population oscillations.
    (E) survivorship curves.

11. Thomas Malthus believed that
    (A) war will prevent humans from reaching carrying capacity.
    (B) the human population size will eventually exceed carrying capacity.
    (C) technological advances will continue to increase human carrying capacity.
    (D) the food supply can keep up with human population growth.
    (E) there are no limits to human population growth.

12. If a population of 10,000 has 300 births, 200 deaths, 50 immigrants and 60 emigrants, what is the populating growth rate?
    (A) 0.9 percent
    (B) 9 percent
    (C) 90 percent
    (D) 2.4 percent
    (E) 24 percent

13. If a country's population growth rate is 5 percent, what is the country's doubling time?
    (A) 5 years
    (B) 35 years
    (C) 14 years
    (D) 42 years
    (E) 72 years

14. Which factor accounts for changes in a country's population growth as is becomes more developed?
    (A) Lowering the total fertility rate
    (B) Increasing life expectancy
    (C) Better care for the elderly
    (D) Availability of medicine
    (E) Reduced emigration

15. Which of the following do we need to determine the population growth rate for a single nation?
    I. Crude birth rate and crude death rate
    II. Doubling time
    III. Immigration and emigration

    (A) I only
    (B) I and II
    (C) II only
    (D) II and III
    (E) I and III

16. What are two reasons for the rapid growth of the human population over the past 8,000 years?
    (A) Lower infant mortality and lower child mortality
    (B) Advances in medicine and technology
    (C) Advances in technology and climate change
    (D) The advent of agriculture and better transportation
    (E) Lack of reliable birth control and increasing urban poverty

Use the following graphs to answer questions 17 and 18.

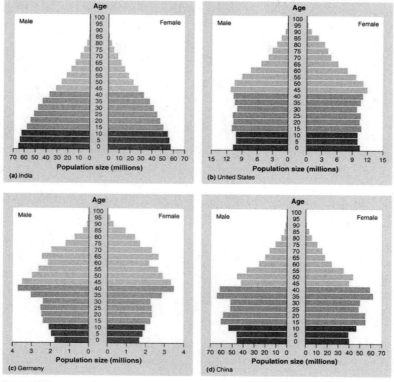

Figure 22.8
Environmental Science for AP®, Second Edition
Data from http://www.census.gov/ipc/www/idb/pyramids.html

17. Which country's age structure diagram shows the most stability?
(A) India
(B) China
(C) Germany
(D) The United States
(E) Germany and China both show population stability

18. Examining India's population structure, we see that
(A) it is growing rapidly.
(B) it is industrialized.
(C) its population momentum is slowing down.
(D) it has a large immigration rate.
(E) its medical care is advanced.

Use the following graph to answer questions 19 and 20.

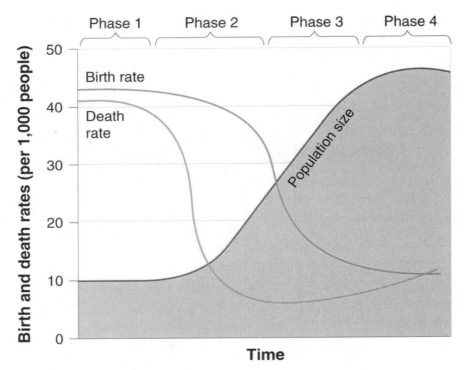

Figure 23.1
Environmental Science for AP®, Second Edition
© 2015 W.H. Freeman and Company

19. When is the population size stable?
   (A) Phase 1
   (B) Phases 1 and 2
   (C) Phase 3
   (D) Phase 4
   (E) Phases 1 and 4

20. When is the population growing most rapidly?
   (A) Phase 1
   (B) Phase 2
   (C) Phase 3
   (D) Phase 4
   (E) Phase 1 and 2

21. If a developing nation quickly reduces its growth rate to 0 percent, its population would
   (A) decrease rapidly.
   (B) decrease slowly.
   (C) level off.
   (D) continue growing for many years then level off.
   (E) grow exponentially.

22. A country with a large, relatively affluent population that is technologically advanced will most likely
    (A) grow exponentially.
    (B) need less medical services.
    (C) have a large environmental impact.
    (D) have a low GDP.
    (E) have a high emigration rate.

23. If a country as a negative net migration rate and low total fertility rate, what can we say about its population?
    (A) Its population is increasing.
    (B) Its population is falling.
    (C) Its population might continue to increase but over a longer time, it will fall.
    (D) Its population will fall but then rebound and increase.
    (E) The country's doubling time is going down.

24. At present, the size of Earth's human population is closest to
    (A) 300 million.
    (B) 1 billion.
    (C) 3.5 billion.
    (D) 7 billion.
    (E) 300 billion.

25. Which is NOT considered in gross domestic product?
    (A) Consumer spending.
    (B) Population growth.
    (C) Investments.
    (D) Government spending.
    (E) Exports minus imports.

26. Survivorship curves represent the distinct patterns of species survival as a function of age. Which correctly pairs type of survivorship curve with the reproductive strategy?
    (A) Type I: r-selected species
    (B) Type II: r-selected species
    (C) Type II: r-selected and K-selected species
    (D) Type I: K-selected species
    (E) Type III: K-selected species

27. Which is NOT associated with a country entering phase 3 of the demographic transition?
    (A) The economy and education systems improve.
    (B) People have fewer children and population numbers stabilize.
    (C) CDR remains low while CBR remains high.
    (D) Resource use increases because of increased affluence.
    (E) TFR decreases and life expectancy increases.

28. Environmental scientists often gauge a country's impact on the environment by evaluating its GDP. Which statement regarding GDP is true?

   (A) Countries with the lowest GDP have the highest pollution levels.

   (B) GDP is primarily determined by government spending.

   (C) A country's ability to enforce environmental regulations goes up with GDP.

   (D) GDP is calculated by multiplying population size by affluence.

   (E) Developing countries typically have a high GDP.

Chapter ⑧ Earth Systems

Chapter Summary

This chapter covers mineral resources, geologic processes, weathering, and soil science. Plate boundaries increase seismic activity and drive the formation of volcanoes and mountain ranges. The rock cycle includes the processes of weathering and erosion. The rock cycle also distributes resources on the surface of Earth. Soil, which is essential for life on Earth, has various physical, chemical, and biological properties. The chapter concludes with a comprehensive discussion of mining processes that remove valuable mineral resources and often harm the environment. Recent legislation has sought to minimize damage, and ongoing reclamation efforts to repair damage from mining have seen some success. The chapter consists of 2 modules:

- **Module 24:** Mineral Resources and Geology
- **Module 25:** Weathering and Soil Science

Chapter Opening Case: *Are Hybrid Electric Vehicles as Environmentally Friendly as We Think?*

This case asks you to consider the environmental impacts of hybrid electric vehicles and all-electric vehicles. Because these vehicles allow us to reduce our use of fossil fuels, many believe they are the answer to our growing transportation needs. However, these vehicles also have environmental impacts. For example, there is a limited supply of metals needed to manufacture batteries, and extracting these metals through mining creates acid mine drainage and fragments habitats.

Do the Math

This chapter contains the following "Do the Math" box to help prepare you for calculation questions you might encounter on the exam.

- "Plate Movement" (page 268)

To make sure you understand the concepts and techniques presented in this box, do the practice problems presented in the text as well as the additional "Practice the Math" problems that appear in Module 24 of this study guide.

Module 24: Environmental Science

BEFORE YOU READ THE MODULE

Focus on Learning Objectives

Use the module learning objectives to guide your reading. On a separate piece of paper, write down each objective and take notes to help you meet each learning objective. After studying this module, you should be able to:

- describe the formation of Earth and the distribution of critical elements on Earth.
- define the theory of plate tectonics and discuss its relevance to the study of the environment.
- describe the rock cycle and discuss its importance in environmental science.

Preview Key Terms

In a notebook or on a separate sheet of paper, create a table like the one shown here to help with learning new key terms in the module. Before you read, fill out the "Prediction" column. Write what you think the term might mean or what it makes you think about. Use examples from your everyday life. There are no wrong answers!

Key Term	Prediction	Definition
Write key term here.	*Write what you think the term means in this column.*	*Define the term here. Add an example and use it in a sentence.*

Key Terms

Core
Mantle
Magma
Asthenosphere
Lithosphere
Crust
Hot spot
Plate tectonics
Tectonic cycle
Subduction

Volcano
Divergent plate boundary
Seafloor spreading
Convergent plate boundary
Transform fault boundary
Fault
Seismic activity
Fault zone
Earthquake
Epicenter

Richter scale
Rock cycle
Igneous rock
Intrusive igneous rock
Extrusive igneous rock
Fracture
Sedimentary rock
Metamorphic rock

Define Key Terms

When you come across a new key term while reading the module, copy the definition into the "Definition" column of your key terms table. Add an example and use the term in a sentence. Compare your initial ideas to the actual definition.

Practice the Math: Plate Movement

Read "Do the Math: Plate Movement," on page 268. Try Your Turn." For more math practice, do the following problems. Remember to show your work.

1. New Hanover is approximately 500 km north of Shaky Acres. If the plate under Shaky Acres is moving at 15mm per year toward New Hanover, how long will it take for Shaky Acres to be located next to New Hanover? [Formula: time = distance ÷ rate]

2. Palm Springs is 718km (446 miles) southwest of San Jose. The plate under Palm Springs is moving northward at about 36 mm per year relative to the plate under San Jose. Given this rate of movement, how long will it take for Palm Springs to be located next to San Jose?

AFTER YOU READ THE MODULE

Review Key Terms

Match the key terms on the left with the definitions on the right.

_____1. Core	a. A vent in the surface of Earth that emits ash, gases, or molten lava
_____2. Mantle	b. A large expanse of rock where a fault has occurred
_____3. Magma	c. Igneous rock that forms when magma rises up and cools in a place underground

_____4. Asthenosphere

d. The exact point on the surface of Earth directly above the location where rock ruptures during an earthquake

_____5. Lithosphere

e. In geology, a place where molten material from Earth's mantle reaches the lithosphere

_____6. Crust

f. A scale that measures the largest ground movement that occurs during an earthquake

_____7. Hot spot

g. Rock formed directly from magma

_____8. Plate tectonics

h. The theory that the lithosphere of Earth is divided into plates, most of which are in constant motion

_____9. Tectonic cycle

i. The innermost zone of Earth's interior, composed mostly of iron and nickel. It includes a liquid outer layer and a solid inner layer

_____10. Subduction

j. The outermost layer of Earth, including the mantle and crust

_____11. Volcano

k. Rock that forms when sediments such as muds, sands, or gravels are compressed by overlying sediments

_____12. Divergent plate boundary

l. Molten rock

_____13. Seafloor spreading

m. Rock that forms when sedimentary rock, igneous rock, or other metamorphic rock is subjected to high temperature and pressure

_____14. Convergent plate boundary

n. The formation of new ocean crust as a result of magma pushing upward and outward from Earth's mantle to the surface

_____15. Transform fault boundary

o. A fracture in rock caused by a movement of Earth's crust

_____16. Fault

p. The layer of Earth located in the outer part of the mantle, composed of semi-molten rock

_____17. Seismic activity

q. The sudden movement of Earth's crust caused by a release of potential energy along a geologic fault and usually causing a vibration or trembling at Earth's surface

_____18. Fault zone

r. The process of one crustal plate passing under another

_____19. Earthquake

s. Rock that forms when magma cools above the surface of Earth

_____20. Epicenter

t. An area where plates move toward one another and collide

_____21. Richter scale

u. The sum of the processes that build up and break down the lithosphere

_____22. Rock cycle

v. In geology, a crack that occurs in rock as it cools

_____23. Igneous rock

w. The layer of Earth above the core, containing magma

_____24. Intrusive igneous rock

x. The frequency and intensity of earthquakes experienced over time

_____25. Extrusive igneous rock

y. The geologic cycle governing the constant formation, alteration, and destruction of rock material that results from tectonics, weathering, and erosion, among other processes

_____26. Fracture

z. In geology, the chemically distinct outermost layer of the lithosphere

_____27. Sedimentary rock

aa. An area beneath the ocean where tectonic plates move away from each other

_____28. Metamorphic rock

bb. An area where tectonic plates move sideways past each other

Module 25: Weathering and Soil Science

BEFORE YOU READ THE MODULE

Focus on Learning Objectives

Use the module learning objectives to guide your reading. On a separate piece of paper, write down each objective and take notes to help you meet each learning objective. After studying this module, you should be able to:

- understand how weathering and erosion occur and how they contribute to element cycling and soil formation.
- explain how soil forms and describe its characteristics.
- describe how humans extract elements and minerals and the social and environmental consequences of these activities.

Preview Key Terms

In a notebook or on a separate sheet of paper, create a table like the one shown here to help with learning new key terms in the module. Before you read, fill out the "Prediction" column. Write what you think the term might mean or what it makes you think about. Use examples from your everyday life. There are no wrong answers!

Key Term	Prediction	Definition
Write key term here.	*Write what you think the term means in this column.*	*Define the term here. Add an example and use it in a sentence.*

Key Terms

Physical weathering
Chemical weathering
Acid precipitation
Acid rain
Erosion
Parent material
Soil degradation
Horizon
O horizon
A horizon
Topsoil
E horizon
B horizon
C horizon

Cation exchange capacity (CEC)
Base saturation
Crustal abundance
Ore
Metal
Reserve
Strip mining
Mining spoils
Tailings
Open-pit mining
Mountaintop removal
Placer mining
Subsurface mining

WHILE YOU READ THE MODULE

Define Key Terms

When you come across a new key term while reading the module, copy the definition into the "Definition" column of your key terms table. Add an example and use the term in a sentence. Compare your initial ideas to the actual definition.

Study the Figure

1. Study Figure 25.7, "Soil horizons" on page 279. To make sure you know the soil horizons, identify them on the version below. Describe the characteristics of each.

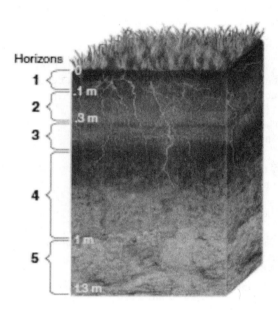

Study the Figure

1. Examine Figure 25.8a, "Soil properties" on page 280. Use the soil texture chart to determine the type of soil if the composition is found to be 85% sand, 15 % silt, and 55% clay.

Review Key Terms

Match the key terms on the left with the definitions on the right.

_____1. Physical weathering

_____2. Chemical weathering

_____3. Acid precipitation (acid rain)

_____4. Erosion

_____5. Parent material

_____6. Soil degradation

_____7. Horizon

_____8. O Horizon

_____9. A horizon (topsoil)

_____10. E horizon

_____11. B horizon

_____12. C. horizon

_____13. Cation exchange capacity (CEC)

_____14. Base saturation

_____15. Crustal abundance

a. A zone of leaching, or eluviation, found in some acidic soils under the O horizon or, less often, the A horizon

b. The breakdown of rocks and minerals by chemical reactions, the dissolving of chemical elements from rocks, or both

c. A concentrated accumulation of minerals from which economically valuable materials can be extracted

d. The ability of a particular soil to absorb and release cations

e. A mining technique in which the entire top of a mountain is removed with explosives

f. The mechanical breakdown of rocks and minerals

g. A soil horizon composed primarily of mineral material with very little organic matter

h. Unwanted waste material created during mining

i. The least-weathered soil horizon, which always occurs beneath the B horizon and is similar to the parent material

j. Frequently the top layer of soil, a zone of organic material and minerals that have been mixed together

k. In resource management, the known quantity of a resource that can be economically recovered

l. Mining techniques used when the desired resource is more than 100 m (328 feet) below the surface of Earth

m. The loss of some or all of a soil's ability to support plant growth

n. The process of looking for minerals, metals, and precious stones in river sediments

o. The physical removal of rock fragments from a landscape or ecosystem

_____16. Ore

p. A mining technique that uses a large visible pit or hole in the ground

_____17. Metal

q. The proportion of soil bases to soil acids, expressed as a percentage

_____18. Reserve

r. The rock material from which the inorganic components of a soil are derived

_____19. Strip mining

s. The organic horizon at the surface of many soils, composed of organic detritus in various stages of decomposition

_____20. Mining spoils (tailings)

t. The average concentration of an element in Earth's crust

_____21. Open-pit mining

u. An element with properties that allow it to conduct electricity and heat energy, and to perform other important functions

_____22. Mountaintop removal

v. Precipitation high in sulfuric acid and nitric acid from reactions between water vapor and sulfur and nitrogen oxides in the atmosphere

_____23. Placer mining

w. The removal of strips of soil and rock to expose ore

_____24. Subsurface mining

x. A horizontal layer in a soil defined by distinctive physical features such as texture and color

Chapter ⑧ Review Exercises

Check Your Understanding

Review "Learning Objectives Revisited" on page 289 of your textbook. Compare the notes you took while reading each module. Complete these exercises to review the chapter.

1. Draw a diagram of Earth's layers.

2. Explain the theory of plate tectonics.

3. Draw a diagram of each of the following:

Divergent Plate Boundary	**Convergent Plate Boundary**	**Transform Fault**

4. Each time you move up the Richter scale you are increasing the damage done by an earthquake by a multiple of _____.

5. Explain the processes that form sedimentary rocks.

6. Diagram the layers of soil.

7. Name and describe the 5 factors that determine soil properties.

8. List three chemical properties and 3 physical properties of soil.

9. Explain in your own words what Figure 25.8 demonstrates.

10. Describe the differences between surface and subsurface mining.

Practice for Free-Response Questions

Valuable mineral and metals are extracted from the Earth's surface. Complete the table below to describe the types of mining processes and evaluate their impact on the environment.

Mining Process	Process Defined	Environmental Impacts
Surface Mining		
Subsurface Mining		

Review and Reflect

Complete these activities to solidify your knowledge of the chapter concepts and key terms. Use a notebook or a separate sheet of paper if necessary.

1. Review your key terms table for each module.

   (a) Which words did you already know? Which were new to you?
   (b) Write a new sentence using each key term.
   (c) Create a set of flash cards that includes each key term. Use the cards to review terms that were new or challenging.
   (d) When you feel comfortable with the new or challenging terms, review all of the cards, including those with familiar terms.
   (e) Save your cards to review before an exam.

2. What are the challenging concepts from this chapter?

   (a) Identify any concepts you found particularly challenging in this chapter.
   (b) Create a list of topics you need to review in preparation for an exam.

3. What questions do you have about concepts in the chapter?

   (a) Note any further questions you might have about material in the chapter.
   (b) Work with a partner to discuss these questions and ask your teacher for help as needed.

4. Write five possible multiple-choice questions based on this chapter. Work with a partner to quiz each other in preparation for an exam.

Chapter (9) Water Resources

Chapter Summary

This chapter demonstrates that fresh water is a critical, scarce resource. The major sources of fresh water are groundwater, surface water, and atmospheric water. When groundwater is extracted faster than it is replaced, the result can be dry water wells and saltwater intrusion. Humans alter the availability of water by constructing levees, dikes, and dams to constrain water movement. Aqueducts transport water. Salt water is desalinized through distillation or reverse osmosis. Developed countries typically have a larger water footprint than developing countries. Most water is used for agriculture, followed by industrial and household use. Recent changes in each of these sectors have improved water use and reuse efficiency and conservation. The chapter consists of 3 modules:

- **Module 26:** The Availability of Water
- **Module 27:** Human Alteration of Water Availability
- **Module 28:** Human Use of Water Now and in the Future

Chapter Opening Case: *Dams and Salmon on the Klamath River*

This opening case provides an excellent opportunity to introduce the importance of water resources to human interests and to ecosystem stability. This case describes the historic use of the Klamath River, which extends from Oregon to California. It has been used over the years as a salmon fishery, a source of irrigation, and a generator of hydroelectric power. For 300 generations it has provided salmon, a traditional food source for Native Americans. Water diversion, overuse, and climate change have harmed the river ecosystem in a variety of ways. In 2009, conservationists, government agencies, the hydroelectric company, and local farmers reached an agreement to ensure that the Klamath River would continue to flow and provide a habitat for salmon populations. An increase in conflicts over water resources around the world makes this case a good model for resolving conflicts in the future.

Do the Math

This chapter contains the following "Do the Math" box to help prepare you for calculation questions you might encounter on the exam.

- "Selecting the Best Washing Machine" (page 314)

To make sure you understand the concepts and techniques presented in this box, do the practice problems presented in the text as well as the additional "Practice the Math" problems that appear in Module 28 of this study guide.

Module 26: Environmental Science

Focus on Learning Objectives

Use the module learning objectives to guide your reading. On a separate piece of paper, write down each objective and take notes to help you meet each learning objective. After studying this module, you should be able to:

- describe the major sources of groundwater.
- identify some of the largest sources of fresh surface water.
- explain the effects of unusually high and low amounts of precipitation.

Preview Key Terms

In a notebook or on a separate sheet of paper, create a table like the one shown here to help with learning new key terms in the module. Before you read, fill out the "Prediction" column. Write what you think the term might mean or what it makes you think about. Use examples from your everyday life. There are no wrong answers!

Key Term	Prediction	Definition
Write key term here.	*Write what you think the term means in this column.*	*Define the term here. Add an example and use it in a sentence.*

Key Terms

Confined aquifer   Spring   Saltwater intrusion

Water table   Artesian well   Floodplain

Groundwater recharge   Cone of depression   Impermeable surface

WHILE YOU READ THE MODULE

Define Key Terms

When you come across a new key term while reading the module, copy the definition into the "Definition" column of your key terms table. Add an example and use the term in a sentence. Compare your initial ideas to the actual definition.

AFTER YOU READ THE MODULE

Review Key Terms

Match the key terms on the left with the definitions on the right.

_____1. Aquifer

a. The uppermost level at which the water in a given area fully saturates rock or soil

_____2. Unconfined aquifer

b. An aquifer surrounded by a layer of impermeable rock or clay that impedes water flow

_____3. Confined aquifer

c. An area lacking groundwater due to rapid withdrawal by a well

_____4. Water table

d. A natural source of water formed when water from an aquifer percolates up to the ground surface

_____5. Groundwater recharge

e. A process by which water percolates through the soil and works its way into an aquifer

_____6. Spring

f. An aquifer made of porous rock covered by soil out of which water can easily flow

_____7. Artesian well

g. The land adjacent to a river

_____8. Cone of depression

h. A well created by drilling a hole into a confined aquifer

_____9. Saltwater intrusion

i. Pavement or buildings that do not allow water penetration

_____10. Floodplain

j. A permeable layer of rock and sediment that contains groundwater

_____11. Impermeable surface

k. An infiltration of salt water in an area where groundwater pressure has been reduced from extensive drilling of wells

Module 27: Human Alteration of Water Availability

BEFORE YOU READ THE MODULE

Focus on Learning Objectives

Use the module learning objectives to guide your reading. On a separate piece of paper, write down each objective and take notes to help you meet each learning objective. After studying this module, you should be able to:

- compare and contrast the roles of levees and dikes.
- explain the benefits and costs of building dams.
- explain the benefits and costs of building aqueducts.
- describe the processes used to convert salt water to fresh water.

Preview Key Terms

In a notebook or on a separate sheet of paper, create a table like the one shown here to help with learning new key terms in the module. Before you read, fill out the "Prediction" column. Write what you think the term might mean or what it makes you think about. Use examples from your everyday life. There are no wrong answers!

Key Term	Prediction	Definition
Write key term here.	*Write what you think the term means in this column.*	*Define the term here. Add an example and use it in a sentence.*

Key Terms

Levee
Dike
Dam
Distillation

Reservoir
Fish ladder
Aqueduct
Reverse osmosis

Desalination
Desalinization

WHILE YOU READ THE MODULE

Define Key Terms

When you come across a new key term while reading the module, copy the definition into the "Definition" column of your key terms table. Add an example and use the term in a sentence. Compare your initial ideas to the actual definition.

Study the Figure

Study Figure 27.5, "Consequences of river diversion" on page 305 and answer the following question.

1. Describe the ecological impacts that have occurred because of water diversion in the Aral Sea.

Study the Figure

Study Figure 27.6, "Desalination technologies" on page 306 and answer the following questions.

1. Describe the desalinization practices of distillation and reverse osmosis.

2. Identify the drawbacks associated with each process.

AFTER YOU READ THE MODULE

Review Key Terms

Match the key terms on the left with the definitions on the right.

_____1. Levee   a. A stair-like structure that allows migrating fish to get around a dam

_____2. Dike   b. A canal or ditch used to carry water from one location to another

_____3. Dam   c. The water body created by a damming a river or stream

_____4. Reservoir   d. The process of removing the salt from salt water

_____5. Fish ladder   e. An enlarged bank built up on each side of a river

_____6. Aqueduct   f. A process of desalination in which water is boiled and the resulting steam is captured and condensed to yield pure water

_____7. Desalination (desalinization) g. A structure built to prevent ocean waters from flooding adjacent land

_____8. Distillation   h. A process of desalination in which water is forced through a thin semipermeable membrane at high pressure

_____9. Reverse osmosis   i. A barrier that runs across a river or stream to control the flow of water

Module 28: Human Use of Water Now and in the Future

BEFORE YOU READ THE MODULE

Focus on Learning Objectives

Use the module learning objectives to guide your reading. On a separate piece of paper, write down each objective and take notes to help you meet each learning objective. After studying this module, you should be able to:

- compare and contrast the four methods of agricultural irrigation.
- describe the major industrial and household uses of water.
- discuss how water ownership and water conservation are important in determining future water availability.

Preview Key Terms

In a notebook or on a separate sheet of paper, create a table like the one shown here to help with learning new key terms in the module. Before you read, fill out the "Prediction" column. Write what you think the term might mean or what it makes you think about. Use examples from your everyday life. There are no wrong answers!

Key Term	Prediction	Definition
Write key term here.	*Write what you think the term means in this column.*	*Define the term here. Add an example and use it in a sentence.*

Key Terms

Water footprint
Hydroponic agriculture

Gray water
Contaminated water

WHILE YOU READ THE MODULE

Define Key Terms

When you come across a new key term while reading the module, copy the definition into the "Definition" column of your key terms table. Add an example and use the term in a sentence. Compare your initial ideas to the actual definition.

Practice the Math: Selecting the Best Washing Machine

Read "Do the Math: Selecting the Best Washing Machine" on page 314. Try "Your Turn." For more math practice, do the following problems. Remember to show your work.

1. You have just moved into a new house and need to purchase a washing machine. Front-loading machines use less water but that they are more expensive. A top-loading machine costs $500 and a new front-loading washing machine costs $1,000.

    (a) If you wash 8 loads of laundry each week, how many loads do you wash each month (assume 4 weeks/month)?

    (b) If the top-loading machine uses 200L of water per load and the front-loading machine uses 100 L of water per load, how many liters of water would the front-loading washing machine save per month?

    (c) If water costs $0.35 for every 1,000L, how much money would be saved each month by using the front-loading washing machine?

2. You have just moved into a new house and need to purchase a new dishwasher.

    (a) If the average family runs the dishwasher 150 times per year, and a traditional dishwasher uses 35 liters more water with each use than a new water-efficient model, how many liters of water would a new water-efficient model save per year?

    (b) If the average cost of water in the United States is $0.50 for every 1000 L, how much money would you save annually by using the more efficient dishwasher?

    (c) If a traditional dishwasher costs $250, whereas the water efficient dishwasher costs $500, how many years will it take for your water savings to pay for the more expensive, high-efficiency dishwasher?

Review Key Terms

Match the key terms on the left with the definitions on the right.

_____1. Water footprint

a. Wastewater from baths, showers, bathrooms, and washing machines

_____2. Hydroponic agriculture

b. The cultivation of plants in greenhouse conditions by immersing roots in a nutrient-rich solution

_____3. Gray water

c. The total daily per capita use of fresh water

_____4. Contaminated water

d. Wastewater from toilets, kitchen sinks, and dishwashers

Chapter (9) Review Exercises

Check Your Understanding

Review "Learning Objectives Revisited" on page 317 of your textbook. Compare the notes you took while reading each module. Complete these exercises to review the chapter.

1. Explain the difference between a confined and an unconfined aquifer.

2. What is saltwater intrusion and what causes this problem?

3. Describe the ecological benefits of freshwater wetlands.

4. Describe some human activities that have contributed to or caused flooding.

5. Describe the benefits and negative consequences of dams.

6. What are the three major uses of fresh water in the world and what percentage of water use goes to each?

Practice for Free-Response Questions

Complete this exercise to build and practice the skills you will need to answer free-response questions on the exam.

Describe four ways in which humans have altered the availability of water and identify environmental consequences of each.

Review and Reflect

Complete these activities to solidify your knowledge of the chapter concepts and key terms. Use a notebook or a separate sheet of paper if necessary.

1. Review your key terms table for each module.

    (a) Which words did you already know? Which were new to you?
    (b) Write a new sentence using each key term.
    (c) Create a set of flash cards that includes each key term. Use the cards to review terms that were new or challenging.
    (d) When you feel comfortable with the new or challenging terms, review all of the cards, including those with familiar terms.
    (e) Save your cards to review before an exam.

2. What are the challenging concepts from this chapter?

    (a) Identify any concepts you found particularly challenging in this chapter.
    (b) Create a list of topics you need to review in preparation for an exam.

3. What questions do you have about concepts in the chapter?

    (a) Note any further questions you might have about material in the chapter.
    (b) Work with a partner to discuss these questions and ask your teacher for help as needed.

4. Write five possible multiple-choice questions based on this chapter. Work with a partner to quiz each other in preparation for an exam.

Unit 4 Multiple-Choice Review Exam

Choose the best answer.

1. Earth's plates are in constant motion because
   (A) pressure from heat and gas in Earth's core pushes plates sideways.
   (B) convection in Earth's mantle causes oceanic plates to spread.
   (C) the density of the lithosphere causes plates to sink.
   (D) hot spots create pressure that shifts plates.
   (E) subduction zones near Japan create fissures that other plates move to fill.

2. Which is associated with seafloor spreading?
   (A) Divergent plate boundaries
   (B) Convergent plate boundaries
   (C) Transform fault boundaries
   (D) Subduction zones
   (E) Hotspots

3. Measured on the Richter scale, an earthquake with a magnitude of 6.0 is how many times greater than an earthquake with a magnitude of 3.0?
   (A) 10
   (B) 100
   (C) 1,000
   (D) 10,000
   (E) 100,000

4. Volcanoes are likely to be found in an area where tectonic plates are
   (A) diverging.
   (B) converging.
   (C) spreading.
   (D) moving sideways past each other.
   (E) stable.

5. Plate movement occurs in which part of Earth's crust?
   (A) Core
   (B) Upper mantle
   (C) Asthenosphere
   (D) Lithosphere
   (E) Lower mantle

6. Worldwide, earthquakes typically occur
   (A) every few years.
   (B) once a year.
   (C) every decade.
   (D) several times each year.
   (E) many times each day.

7. Which does NOT contribute to the formation of sedimentary rock?
   (A) Weathering
   (B) Erosion
   (C) Transportation
   (D) Compression
   (E) Melting

8. Fossils are most likely to be found in
   (A) igneous rock.
   (B) metamorphic rock.
   (C) sedimentary rock.
   (D) basalt.
   (E) granite.

9. Acid precipitation can cause
   (A) physical weathering.
   (B) erosion.
   (C) sedimentation.
   (D) an interruption in the tectonic cycle.
   (E) chemical weathering.

10. Which type of soil allows the most water to infiltrate?
    (A) Sand
    (B) Silt
    (C) Clay
    (D) Humus
    (E) Loam

11. Soil found in tropical rain forests is generally
    (A) rich in organic material.
    (B) rich in quartz sand.
    (C) deep and porous.
    (D) nutrient poor.
    (E) acidic.

12. The soil layer most similar to the parent material is the
    (A) O horizon.
    (B) A horizon.
    (C) B horizon.
    (D) C horizon.
    (E) E horizon.

13. When rocks are subjected to extreme heat and pressure what type of rock is formed?
    (A) Igneous extrusive
    (B) Igneous intrusive
    (C) Metamorphic
    (D) Sedimentary
    (E) Granitic

14. Which does NOT contribute to soil degradation?
    (A) Machines
    (B) Humans
    (C) Livestock
    (D) Plowing
    (E) Composting

15. The Surface Mining Control and Reclamation Act of 1977 regulates
    (A) compensation for mine workers who are harmed by coal dust.
    (B) surface mining of coal and subsurface effects of surface mining.
    (C) how much water a mine can use.
    (D) the type of fill material that must be used when reclaiming a mine.
    (E) the equipment used in surface mining.

16. Ice and glaciers represent what percentage of the freshwater on Earth?
    (A) 22 percent
    (B) 0.5 percent
    (C) 97 percent
    (D) 3 percent
    (E) 77 percent

17. Confined aquifers
    (A) are polluted more easily than unconfined aquifers.
    (B) are easily recharged.
    (C) are covered by impermeable rock.
    (D) are accessed only through pumping.
    (E) are covered by porous rock and soil.

18. The Ogallala aquifer is
    (A) the largest aquifer in the United States.
    (B) the largest aquifer in the world.
    (C) heavily polluted.
    (D) below the water table.
    (E) contaminated by saltwater intrusion.

19. Water diversion has split which body of water and reduced its surface area by 60 percent?
    (A) Lake Victoria
    (B) The Caspian Sea
    (C) Lake Baikal
    (D) The Aral Sea
    (E) Mono Lake.

20. Which part of the world produces 50 percent of the world's desalinated water?
    (A) North America
    (B) Middle East
    (C) Europe
    (D) Asia
    (E) South America

21. What is the purpose of a fish ladder?
    (A) They allow migrating fish to get around a dam.
    (B) They serve as a location for breeding.
    (C) They allow people to transport fish.
    (D) They prevent invasive species from entering a reservoir.
    (E) They save fish when a levee breaks.

22. Over-pumping an aquifer can cause all of the following except
    (A) saltwater intrusion.
    (B) cone of depression.
    (C) eutrophication.
    (D) dry wells.
    (E) land and property loss.

23. Desalinization is
    (A) used extensively across Europe.
    (B) causing drought.
    (C) a problem with saltwater intrusion.
    (D) occurring at a rapid rate due to over irrigating farmland.
    (E) helping water-poor countries obtain fresh water.

24. Which type of irrigation is over 95 percent efficient?
    (A) Spray irrigation
    (B) Flood irrigation
    (C) Furrow irrigation
    (D) Drip irrigation
    (E) Hydroponic irrigation

25. Which activity accounts for the largest percentage of indoor household water use in the United States?
    (A) Toilet flushing
    (B) Bathing
    (C) Laundry
    (D) Cooking.
    (E) Drinking

26. Which processes are involved in the formation of sedimentary rock?
    (A) Subduction and melting
    (B) Cooling and crystallization
    (C) Uplift and heat
    (D) Heat and pressure
    (E) Erosion, transport and compression

27. Desalination of water could provide fresh water to water-poor regions. Which of the following are costs associated with desalination techniques including distillation and reverse osmosis?

   I. Desalination systems require large investments to build, maintain and repair.

   II. Brine is potentially harmful to plant and animal life and could contaminate soil and ocean waters.

   III. Both processes require water to be boiled which is energy intensive.

(A) I only

(B) II only

(C) I and II

(D) I and III

(E) I, II and III

Chapter (10) Land, Public and Private

Chapter Summary

This chapter introduces issues associated with land use, issues that often create conflict. This chapter explores how our use of land affects the environment and what we can do to minimize negative impacts. It examines land use and management in both private and public sectors, as well as sustainable land use practices. The chapter also covers urban sprawl and the problems it creates. Finally, it describes smart growth as a solution to urban sprawl.
The chapter consists of 2 modules:

- **Module 29:** Land Use Concepts and Classification
- **Module 30:** Land Management Practices

Chapter Opening Case: *Who Owns a Tree? Julia Butterfly Hill versus Maxxam*

The chapter opening case introduces you to the practice of timber harvesting. It discusses two types of timber harvesting practices: clear-cutting and selective cutting. It describes an event that occurred on privately owned land in a Stafford, California, redwood forest. A company planned to clear-cut the forest, but activist Julia Butterfly Hill took action to protect a single tree.

Module 29: Environmental Science

BEFORE YOU READ THE MODULE

Focus on Learning Objectives

Use the module learning objectives to guide your reading. On a separate piece of paper, write down each objective and take notes to help you meet each learning objective. After studying this module, you should be able to:

- explain how human land use affects the environment.
- describe the various categories of public land used globally and in the United States.

Preview Key Terms

In a notebook or on a separate sheet of paper, create a table like the one shown here to help with learning new key terms in the module. Before you read, fill out the "Prediction" column. Write what you think the term might mean or what it makes you think about. Use examples from your everyday life. There are no wrong answers!

Key Term	Prediction	Definition
Write key term here.	*Write what you think the term means in this column.*	*Define the term here. Add an example and use it in a sentence.*

Key Terms

Tragedy of the commons
Externality
Maximum sustainable yield (MSY)

Resource conservation ethic
Multiple-use lands

WHILE YOU READ THE MODULE

Define Key Terms

When you come across a new key term while reading the module, copy the definition into the "Definition" column of your key terms table. Add an example and use the term in a sentence. Compare your initial ideas to the actual definition.

Study the Figure

Examine Figure 29.3, "Maximum sustainable yield" on page 334 and answer the following question.

1. Why is maximum sustainable yield located where a population reaches approximately half the carrying capacity?

AFTER YOU READ THE MODULE

Review Key Terms

Match the key terms on the left with the definitions on the right.

_____1. Tragedy of the commons

a. The tendency of a shared, limited resource to become depleted because people act from self-interest for short-term gain

_____2. Externality

b. The belief that people should maximize use of resources, based on the greatest good for everyone

_____3. Maximum sustainable yield (MSY)

c. The maximum amount of a renewable resource that can be harvested without compromising the future availability of that resource

_____4. Resource conservation ethic

d. A U.S. classification used to designate lands that may be used for recreation, grazing, timber harvesting, and mineral extraction

_____5. Multiple-use lands

e. The cost or benefit of a good or service that is not included in the purchase price of that good or service

Module 30: Land Management Practices

BEFORE YOU READ THE MODULE

Focus on Learning Objectives

Use the module learning objectives to guide your reading. On a separate piece of paper, write down each objective and take notes to help you meet each learning objective. After studying this module, you should be able to:

- explain specific land management practices for rangelands and forests.
- describe contemporary problems in residential land use and some potential solutions.

Preview Key Terms

In a notebook or on a separate sheet of paper, create a table like the one shown here to help with learning new key terms in the module. Before you read, fill out the "Prediction" column. Write what you think the term might mean or what it makes you think about. Use examples from your everyday life. There are no wrong answers!

Key Term	Prediction	Definition
Write key term here.	*Write what you think the term means in this column.*	*Define the term here. Add an example and use it in a sentence.*

Key Terms

Rangeland
Forest
Clear-cutting
Selective cutting
Ecologically sustainable forestry
Tree plantation
Prescribed burn
National wildlife refuge
National wilderness area
National Environmental Policy Act (NEPA)
Environmental impact statement (EIS)
Environmental mitigation plan
Endangered Species Act
Suburban

Exurban
Urban sprawl
Urban blight
Highway Trust Fund
Induced demand
Zoning
Multi-use zoning
Smart growth
Stakeholder
Sense of place
Transit-oriented development (TOD)
Infill
Urban growth boundaries
Eminent domain

WHILE YOU READ THE MODULE

Define Key Terms

When you come across a new key term while reading the module, copy the definition into the "Definition" column of your key terms table. Add an example and use the term in a sentence. Compare your initial ideas to the actual definition.

Study the Figure

1. Using Figure 30.2, "Timber harvest practices" on page 340, describe the timber harvest method of clear cutting. What are the benefits and disadvantages?

2. Using Figure 30.2, "Timber harvest practices" on page 340, describe the timber harvest method of selective cutting. What are the benefits and disadvantages?

Review Key Terms

Match the key terms on the left with the definitions on the right.

_____1. Rangeland

a. A plan that outlines how a developer will address concerns raised by a project's impact on the environment

_____2. Forest

b. A 1969 U.S. federal act that mandates an environmental assessment of all projects involving federal money or federal permits

_____3. Clear-cutting

c. An approach to removing trees from forests in ways that do not unduly affect the viability of other trees

_____4. Selective cutting

d. A person or organization with an interest in a particular place or issue

_____5. Ecologically sustainable forestry

e. The feeling that an area has a distinct and meaningful character

_____6. Tree plantation

f. An area set aside with the intent of preserving a large tract of intact ecosystem or a landscape

_____7. Prescribed burn

g. Development that attempts to focus dense residential and retail development around stops for public transportation, a component of smart growth

_____8. National wildlife refuge

h. A large area typically planted with a single rapidly growing tree species

_____9. National wilderness area

i. Development that fills in vacant lots within existing communities

_____10. National Environmental Policy Act (NEPA)

j. Land dominated by trees and other woody vegetation and sometimes used for commercial logging

_____11. Environmental impact statement (EIS)

k. A restriction on development outside a designated area

_____12. Environmental mitigation plan

l. A principle that grants government the power to acquire a property at fair market value even if the owner does not wish to sell it

_____13. Endangered Species Act

m. The method of harvesting trees that involves the removal of single trees or a relatively small number of trees from among many in a forest

_____14. Suburb

n. A fire deliberately set under controlled conditions in order to reduce the accumulation of dead biomass on a forest floor

_____15. Exurb

o. An area similar to a suburb, but unconnected to any central city or densely populated area

_____16. Urban sprawl

p. A method of harvesting trees that involves removing all or almost all of the trees within an area

_____17. Urban blight

q. Urbanized areas that spread into rural areas, removing clear boundaries between the two

_____18. Highway Trust Fund

r. An area surrounding a metropolitan center, with a comparatively low population density

_____19. Induced demand

s. The degradation of the built and social environments of the city that often accompanies and accelerates migration to the suburbs

_____20. Zoning

t. A dry open grassland

_____21. Multi-use zoning

u. A U.S. federal fund that pays for the construction and maintenance of roads and highways

_____22. Smart growth

v. The phenomenon in which an increase in the supply of a good causes demand to grow

_____23. Stakeholder

w. A planning tool used to separate industry and business from residential neighborhoods

_____24. Sense of place

x. A federal public land managed for the primary purpose of protecting wildlife

_____25. Transit-oriented development (TOD)

y. A 1973 U.S. act designed to protect species from extinction

_____26. Infill

z. A document outlining the scope and purpose of a development project, describing the environmental context, suggesting alternative approaches to the project, and analyzing the environmental impact of each alternative

_____27. Urban growth boundary

aa. A zoning classification that allows retail and high-density residential development to coexist in the same area

_____28. Eminent domain

bb. A set of principles for community planning that focuses on strategies to encourage the development of sustainable, healthy communities

Chapter ⑩ Review Exercises

Check Your Understanding

Review "Learning Objectives Revisited" on page 352 of your textbook. Compare the notes you took while reading each module. Complete these exercises to review the chapter.

1. What is the single largest cause of species extinctions today?

2. Give an example of a negative externality.

3. What are some of the uses of public and private land in the United States?

4. Describe the negative effects of fire suppression.

5. Describe urban blight and how it occurs.

Practice for Free-Response Questions

Complete this exercise to build and practice the skills you will need to answer free-response questions on the exam. Use a separate sheet of paper if necessary.

Explain how maximum sustainable yield ensures that a resource is used sustainably yet still provides an economic benefit.

Review and Reflect

Complete these activities to solidify your knowledge of the chapter concepts and key terms. Use a notebook or a separate sheet of paper if necessary.

1. Review your key terms table for each module.

    (a) Which words did you already know? Which were new to you?
    (b) Write a new sentence using each key term.
    (c) Create a set of flash cards that includes each key term. Use the cards to review terms that were new or challenging.
    (d) When you feel comfortable with the new or challenging terms, review all of the cards, including those with familiar terms.
    (e) Save your cards to review before an exam.

2. What are the challenging concepts from this chapter?

    (a) Identify any concepts you found particularly challenging in this chapter.
    (b) Create a list of topics you need to review in preparation for an exam.

3. What questions do you have about concepts in the chapter?

    (a) Note any further questions you might have about material in the chapter.
    (b) Work with a partner to discuss these questions and ask your teacher for help as needed.

4. Write five possible multiple-choice questions based on this chapter. Work with a partner to quiz each other in preparation for an exam.

Chapter (11) Feeding the World

Chapter Summary

This chapter explores how agriculture has developed to meet the needs of a growing population. Technological advances such as inorganic fertilizers, pesticides, and genetically modified crops have helped humans grow increasing amounts of food. However, these advances do not come without some environmental costs. The chapter also discusses alternatives to industrial farming methods. Sustainable agricultural techniques such as no-till, intercropping, crop rotation, contour plowing, and integrated pest management are all discussed in detail. In addition to advances in agriculture, there have also been changes in the way we raise animals and harvest fish; concentrated animal feeding operations (CAFOs) and aquaculture are relatively new methods that have helped increase our food supply. The chapter consists of 3 modules:

- **Module 31:** Human Nutritional Needs
- **Module 32:** Modern Large-Scale Farming Methods
- **Module 33:** Alternatives to Industrial Farming Methods

Chapter Opening Case: *A Farm Where Animals Do Most of the Work*

This opening case points out that using cow and chicken manure is a good way to recycle nutrients back into the soil. Since nitrogen is one of the major components of fertilizer, this would be a good time to review the nitrogen cycle. Have your students draw the nitrogen cycle starting with atmospheric nitrogen.

Do the Math

This chapter contains the following "Do the Math" box to help prepare you for calculation questions you might encounter on the exam.

- " Land Needed for Food" (page 365)

To make sure you understand the concepts and techniques presented in this box, do the practice problems presented in the text as well as the additional "Practice the Math" problems that appear in Module 32 of this study guide.

Module 31: Human Nutritional Needs

BEFORE YOU READ THE MODULE

Focus on Learning Objectives

Use the module learning objectives to guide your reading. On a separate piece of paper, write down each objective and take notes to help you meet each learning objective. After studying this module, you should be able to:

- describe human nutritional requirements.
- explain why nutritional requirements are not being met in various parts of the world.

Preview Key Terms

In a notebook or on a separate sheet of paper, create a table like the one shown here to help with learning new key terms in the module. Before you read, fill out the "Prediction" column. Write what you think the term might mean or what it makes you think about. Use examples from your everyday life. There are no wrong answers!

Key Term	Prediction	Definition
Write key term here.	*Write what you think the term means in this column.*	*Define the term here. Add an example and use it in a sentence.*

Key Terms

Undernutrition	Food insecurity	Overnutrition
Malnourished	Famine	Meat
Food security	Anemia	

WHILE YOU READ THE MODULE

Define Key Terms

When you come across a new key term while reading the module, copy the definition into the "Definition" column of your key terms table. Add an example and use the term in a sentence. Compare your initial ideas to the actual definition.

Review Key Terms

Match the key terms on the left with the definitions on the right.

_____1. Undernutrition

a. A deficiency of iron

_____2. Malnourished

b. Having a diet that lacks the correct balance of proteins, carbohydrates, vitamins, and minerals

_____3. Food security

c. Ingestion of too many calories and a lack of balance of foods and nutrients

_____4. Food insecurity

d. Livestock or poultry consumed as food

_____5. Famine

e. The condition in which not enough calories are ingested to maintain health

_____6. Anemia

f. A condition in which people do not have adequate access to food.

_____7. Overnutrition

g. A condition in which people have access to sufficient, safe, and nutritious food that meets their dietary needs for an active and healthy life

_____8. Meat

h. The condition in which food insecurity is so extreme that large numbers of deaths occur in a given area over a relatively short period

Module 32: Modern Large-Scale Farming Methods

BEFORE YOU READ THE MODULE

Focus on Learning Objectives

Use the module learning objectives to guide your reading. On a separate piece of paper, write down each objective and take notes to help you meet each learning objective. After studying this module, you should be able to:

- describe modern, large-scale agricultural methods.
- explain the benefits and consequences of genetically modified organisms.
- discuss the large-scale raising of meat and fish.

Preview Key Terms

In a notebook or on a separate sheet of paper, create a table like the one shown here to help with learning new key terms in the module. Before you read, fill out the "Prediction" column. Write what you think the term might mean or what it makes you think about. Use examples from your everyday life. There are no wrong answers!

Key Term	Prediction	Definition
Write key term here.	*Write what you think the term means in this column.*	*Define the term here. Add an example and use it in a sentence.*

Key Terms

Industrial agriculture
Agribusiness
Energy subsidy
Green Revolution
Economies of scale
Waterlogging
Salinization
Organic fertilizer
Synthetic fertilizer
Inorganic fertilizer
Monocropping
Pesticides
Insecticide

Herbicide
Broad-spectrum pesticide
Selective pesticide
Persistent pesticide
Nonpersistent pesticide
Pesticide resistant
Pesticide treadmill
Concentrated animal feeding operation (CAFO)
Fishery
Fishery collapse
Bycatch

WHILE YOU READ THE MODULE

Define Key Terms

When you come across a new key term while reading the module, copy the definition into the "Definition" column of your key terms table. Add an example and use the term in a sentence. Compare your initial ideas to the actual definition.

Practice the Math: Land Needed for Food

Read "Do the Math: Land Needed for Food," on page 365. Try "Your Turn." For more math practice, do the following problems. Remember to show your work. Use a separate sheet of paper if necessary.

1. If a person's food requirement per day is 2,200 kilocalories, how many kilocalories does that person need per month (assume a 30-day month)?

2. If a person only eats apples and each apple contains 53 kilocalories, how many apples will the person need to eat to get all the kilocalories they require per day? Per month?

3. If there are 6.8 billion people on the planet, how many apples would be needed per day to meet all the kilocalorie demands for the world?

4. On farms in the Midwestern United States, a hectare of land yields roughly 40 bushels of wheat. Each bushel of wheat weighs approximately 27 kilograms. One kilogram of wheat provides 3,500 kilocalories. Assume that a person only eats wheat and must consume 2,000 kcal per day. How much land is required to feed that person for one year?

Study the Figure

Examine Figure 32.6, "The pesticide treadmill" on page 370.

1. Determine if the pesticide treadmill is an example of either a positive or negative feedback loop. Explain why stronger pesticides and greater amounts of pesticides must be used over time.

Review Key Terms

Match the key terms on the left with the definitions on the right.

_____1. Industrial agriculture

_____2. Energy subsidy

_____3. Green Revolution

_____4. Economies of scale

_____5. Waterlogging

_____6. Salinization

_____7. Organic fertilizer

_____8. Synthetic fertilizer

_____9. Monocropping

_____10. Pesticide

_____11. Insecticide

_____12. Herbicide

_____13. Broad-spectrum pesticide

_____14. Selective pesticide

_____15. Persistent pesticide

a. A substance, either natural or synthetic, that kills or controls organisms that people consider pests

b. The unintentional catch of nontarget species while fishing

c. Fertilizer produced commercially, normally with the use of fossil fuels. Also known as inorganic fertilizer

d. The decline of a fish population by 90 percent or more

e. The observation that average costs of production fall as output increases

f. A large indoor or outdoor structure designed for maximum output

g. A form of soil degradation that occurs when soil remains under water for prolonged periods

h. A commercially harvestable population of fish within a particular ecological region

i. The fossil fuel energy and human energy input per calorie of food produced

j. A pesticide that targets plant species that compete with crops

k. A cycle of pesticide development, followed by pest resistance, followed by new pesticide development

l. An agricultural method that utilizes large plantings of a single species or variety

m. A shift in agricultural practices in the twentieth century that included new management techniques, mechanization, fertilization, irrigation, and improved crop varieties, and that resulted in increased food output

n. A trait possessed by certain individuals that are exposed to a pesticide and survive

o. A pesticide that breaks down rapidly, usually in weeks or months

_____16. Nonpersistent pesticide

p. A form of soil degradation that occurs when the small amount of salts in irrigation water becomes highly concentrated on the soil surface through evaporation

_____17. Pesticide resistance

q. A pesticide that remains in the environment for a long time

_____18. Pesticide treadmill

r. Fertilizer composed of organic matter from plants and animals

_____19. Concentrated animal feeding operation (CAFO)

s. A pesticide that targets a narrow range of organisms

_____20. Fishery

t. A pesticide that targets species of insects and other invertebrates that consume crops

_____21. Fishery collapse

u. Agriculture that applies the techniques of mechanization and standardization. Also known as agribusiness

_____22. Bycatch

v. A pesticide that kills many different types of pest

Module 33: Scientific Method

BEFORE YOU READ THE MODULE

Focus on Learning Objectives

Use the module learning objectives to guide your reading. On a separate piece of paper, write down each objective and take notes to help you meet each learning objective. After studying this module, you should be able to:

- describe alternatives to conventional farming methods.
- explain alternative techniques used in farming animals and in fishing and aquaculture.

Preview Key Terms

In a notebook or on a separate sheet of paper, create a table like the one shown here to help with learning new key terms in the module. Before you read, fill out the "Prediction" column. Write what you think the term might mean or what it makes you think about. Use examples from your everyday life. There are no wrong answers!

Key Term	Prediction	Definition
Write key term here.	*Write what you think the term means in this column.*	*Define the term here. Add an example and use it in a sentence.*

Key Terms

Shifting agriculture
Desertification
Nomadic grazing
Sustainable agriculture
Intercropping
Crop rotation
Agroforestry
Contour plowing

Perennial plant
Annual plant
No-till agriculture
Integrated pest management (IPM)
Organic agriculture
Individual transferable quota (ITQ)
Aquaculture

WHILE YOU READ THE MODULE

Define Key Terms

When you come across a new key term while reading the module, copy the definition into the "Definition" column of your key terms table. Add an example and use the term in a sentence. Compare your initial ideas to the actual definition.

AFTER YOU READ THE MODULE

Review Key Terms

Match the key terms on the left with the definitions on the right.

_____1. Shifting agriculture

_____2. Desertification

_____3. Nomadic grazing

a. Production of crops without the use of synthetic pesticides or fertilizers

b. Farming aquatic organisms such as fish, shellfish, and seaweeds

c. An agricultural method in which farmers do not turn the soil between seasons as a means of reducing topsoil erosion

_____4. Sustainable agriculture

_____5. Intercropping

_____6. Crop rotation

_____7. Agroforestry

_____8. Contour plowing

_____9. Perennial plant

_____10. Annual plant

_____11. No-till agriculture

_____12. Integrated pest management (IPM)

_____13. Organic agriculture

_____14. Individual transferable quote (ITQ)

_____15. Aquaculture

d. A fishery management program in which individual fishers are given a total allowable catch of fish in a season that they can either catch or sell

e. An agricultural method in which two or more crop species are planted in the same field at the same time to promote a synergistic interaction

f. An agricultural practice that uses a variety of techniques designed to minimize pesticide inputs

g. A plant that lives only one season

h. An agricultural method in which land is cleared and used for a few years until the soil is depleted of nutrients

i. An agricultural technique in which plowing and harvesting are done parallel to the topographic contours of the land

j. The transformation of arable, productive land to desert or unproductive land due to climate change or destructive land use

k. An agricultural technique in which trees and vegetables are intercropped

l. Agriculture that fulfills the need for food and fiber while enhancing the quality of the soil, minimizing the use of nonrenewable resources, and allowing economic viability for the farmer

m. An agricultural technique in which crop species in a field are rotated from season to season

n. The feeding of herds of animals by moving them to seasonally productive feeding grounds, often over long distances

o. A plant that lives for multiple years

Chapter (11) Review Exercises

Check Your Understanding

Review "Learning Objectives Revisited" on page 385 of your textbook. Compare the notes you took while reading each module. Complete these exercises to review the chapter.

1. Distinguish between undernutrition and malnutrition and provide causes of each.

2. Where do most of the energy subsidies in modern agriculture go?

3. How can irrigation contribute to soil degradation?

4. List disadvantages and advantages of using synthetic fertilizers.

5. How can shifting agriculture be detrimental to the habitat of the area?

6. List four sustainable agriculture practices and explain how they differ from traditional practices.

7. What are some environmental and health consequences of high-density animal farming?

Practice for Free-Response Questions

Complete this exercise to build and practice the skills you will need to answer free-response questions on the exam. Use a separate sheet of paper if necessary.

In 1972, DDT (dichlorodiphenyltrichloroethane), a persistent pesticide, was banned in the United States. Describe the effects of this substance and explain why the United States banned it.

Review and Reflect

Complete these activities to solidify your knowledge of the chapter concepts and key terms. Use a notebook or a separate sheet of paper if necessary.

1. Review your key terms table for each module.

    (a) Which words did you already know? Which were new to you?
    (b) Write a new sentence using each key term.
    (c) Create a set of flash cards that includes each key term. Use the cards to review terms that were new or challenging.
    (d) When you feel comfortable with the new or challenging terms, review all of the cards, including those with familiar terms.
    (e) Save your cards to review before an exam.

2. What are the challenging concepts from this chapter?

    (a) Identify any concepts you found particularly challenging in this chapter.
    (b) Create a list of topics you need to review in preparation for an exam.

3. What questions do you have about concepts in the chapter?

    (a) Note any further questions you might have about material in the chapter.
    (b) Work with a partner to discuss these questions and ask your teacher for help as needed.

4. Write five possible multiple-choice questions based on this chapter. Work with a partner to quiz each other in preparation for an exam.

Unit 5 Multiple-Choice Review Exam

Choose the best answer.

1. Which would be an example of the tragedy of the commons?
    (A) Air pollution caused by a factory
    (B) Deforestation on private land
    (C) A school converting its football field into a parking lot
    (D) A farmer draining a lake on his land
    (E) Ranchers grazing cattle on publicly-owned land

2. Which is an example of a positive externality?
    (A) Post-holiday sale prices
    (B) Feeling relaxed after a vacation
    (C) Pollution from automobile exhaust
    (D) An existing mangrove forest preventing coastal storm damage
    (E) Desertification of grazing land because of cattle farming

3. A logging company clears selectively clears a job site and removes only enough large trees to let light reach the younger trees. What is the purpose of this practice?
    (A) To prevent any negative externalities
    (B) To create a positive externality, despite logging trees in the forest
    (C) To avoid a tragedy of the commons
    (D) To harvest using the principle of maximum sustainable yield
    (E) To avoid harming the animals who live in the forest

4. The majority of land in the United States is used for
    (A) residential and commercial property.
    (B) timber production.
    (C) grassland and grazing land.
    (D) forests and grazing land.
    (E) recreational and wildlife land.

5. Which is the most profitable way to harvest trees?
    (A) Selective cutting
    (B) Clear-cutting
    (C) Waterlogging
    (D) Using horses instead of machines
    (E) Prescribed burn

6. Prescribed burns are used to
    (A) destroy invasive species.
    (B) reduce herbicide use.
    (C) increase logging profits.
    (D) clear land economically.
    (E) reduce the accumulation of dead biomass.

7. Which is a concern about tree plantations?
   (A) Cost of harvesting trees
   (B) Lack of biodiversity
   (C) Introduction of invasive species
   (D) Difficulty with fire control
   (E) Loss of jobs in the logging industry

8. Which is NOT a consequence of clear-cutting?
   (A) Warmer water temperatures in nearby water bodies
   (B) Erosion
   (C) Introduction of invasive species
   (D) Loss of soil nutrients
   (E) Sediment buildup in nearby streams

9. Which is an example of urban sprawl?
   (A) A heavily populated downtown business center
   (B) A pasture populated with many cattle
   (C) Rangeland for grazing animals
   (D) Suburbs with expanding neighborhoods
   (E) A large shopping complex in a city railway station

10. Which is a positive effect of traditional zoning laws?
    (A) It promotes automobile-dependent development.
    (B) It does not consider environmental damage from construction.
    (C) It prohibits suburban neighborhoods from developing a main street.
    (D) It addresses issues of safety.
    (E) It is flexible regarding multi-use options.

11. Which is a consequence of urban sprawl?
    (A) Large homes close together
    (B) Farmland close to a river
    (C) Housing and retail shops separated by miles of road
    (D) Decreased traffic congestion
    (E) Decreased gasoline use

12. Which is NOT a characteristic of smart growth?
    (A) Mixed land use
    (B) Walkable neighborhoods
    (C) Compact building design
    (D) A variety of transportation choices
    (E) Separate areas for shopping and residences

13. Building on a vacant lot in an existing community is known as
    (A) urban sprawl.
    (B) urban blight.
    (C) transit-oriented development.
    (D) infill.
    (E) multi-use zoning.

14. According to the World Health Organization (WHO) nearly half of the world's population is
    (A) malnourished.
    (B) anemic.
    (C) overweight.
    (D) eating too much beef.
    (E) experiencing overnutrition.

15. The largest component of the human diet is
    (A) grain.
    (B) meat.
    (C) fat.
    (D) vegetables.
    (E) soy.

16. Which is NOT a benefit of aquaculture?
    (A) It provides a supply of protein for people worldwide.
    (B) It alleviates some of the pressure on over-exploited fisheries.
    (C) It boosts the economies of many developing countries.
    (D) It increases biodiversity.
    (E) It can be practiced in many parts of the world.

17. Which of the following is NOT a practice of the Green Revolution?
    (A) Mechanization
    (B) Overgrazing
    (C) Irrigation
    (D) Monocropping
    (E) Pesticide use

18. The pesticide treadmill occurs when
    (A) a farmer uses biological pest controls and the biological agent reproduces.
    (B) a homeowner uses excessive fertilizer.
    (C) an invasive pest species takes over cropland.
    (D) pests reproduce exponentially.
    (E) a new pesticide must be used because of pest resistance.

19. Which substance kills pests but remains in the environment for a long time?
    (A) Persistent herbicide
    (B) Nonpersistent herbicide
    (C) Persistent insecticide
    (D) Nonpersistent insecticide
    (E) Selective pesticide

20. All of the following are benefits of genetic engineering EXCEPT
    (A) greater yield.
    (B) better food quality.
    (C) less pesticide use.
    (D) higher profits.
    (E) less soil erosion.

21. Which is characteristic of a genetically modified crop species?
   I.   A gene from one organism is transferred to a different organism
   II. It poses no risk to wild plant species
   III. It allows a farmer to increase revenue
   (A) I only
   (B) I and II
   (C) II and III
   (D) I and III
   (E) III only

22. Desertification is happening most rapidly in which country?
   (A) The United States
   (B) Africa
   (C) Europe
   (D) Canada
   (E) Japan

23. Which is NOT a technique included in integrated pest management?
   (A) Crop rotation
   (B) Intercropping
   (C) Using pest-resistant crop varieties
   (D) Encouraging predator insects
   (E) Using persistent pesticides

24. Which is NOT a characteristic of high-density animal farming?
   (A) Increased strains of antibiotic-resistant microorganisms
   (B) Runoff problems into area waterways
   (C) Producing free-range chicken and beef
   (D) Risks to human health
   (E) Minimized land costs to farmers

25. What is one way to alleviate some of the human-caused pressure on overexploited fisheries?
   (A) Increase aquaculture practices
   (B) Use long line fishing practices
   (C) Encourage purse-seine fishing
   (D) Use dragnets
   (E) Increase the consumption of bycatch organisms

Chapter (12) Nonrenewable Energy Resources

Chapter Summary

This chapter explores how energy is used, the variety of nonrenewable resources, and how to project future supplies of these resources. This chapter offers an opportunity for students to focus on the advantages and disadvantages of each type of energy resource. For example, coal is abundant, energy-dense, easy to obtain, and relatively inexpensive. However, coal has impurities that are released into the atmosphere when it is burned, it leaves behind large deposits of ash, and it contributes to the increasing atmospheric concentrations of carbon dioxide, sulfur dioxide, and particulates. The chapter consists of 3 modules:

- **Module 34:** Patterns of Energy Use
- **Module 35:** Fossil Fuel Resources
- **Module 36:** Nuclear Energy Resources

Chapter Opening Case: *All Energy Use Has Consequences*

This opening case discusses some of the risks associated with use of nonrenewable fossil fuels. It mentions both the *Exxon Valdez* spill in 1989 and the BP *Deepwater Horizon* oil rig blowout in 2010, as well as the more recent events at the Fukushima nuclear power plant in Japan. The United States consumes more energy (EJ) per capita than any other country in the world. The technological progress associated with increased energy consumption provides many benefits, but our dependence on fossil fuels for energy is not without long-term costs.

Do the Math

This chapter contains the following "Do the Math" boxes to help prepare you for calculation questions you might encounter on the exam.

- "Efficiency of Travel" (page 404)
- "Calculating Energy Supply" (page 407)
- "Calculating Half-Lives" (page 422)

To make sure you understand the concepts and techniques presented in these boxes, do the practice problems presented in the text as well as the additional "Practice the Math" problems that appear in Module 34 and 36 of this study guide.

Module 34: Patterns of Energy Use

BEFORE YOU READ THE MODULE

Focus on Learning Objectives

Use the module learning objectives to guide your reading. On a separate piece of paper, write down each objective and take notes to help you meet each learning objective. After studying this module, you should be able to:

- describe the use of nonrenewable energy in the world and in the United States.
- explain why different forms of energy are best suited for certain purposes.
- understand the primary ways that electricity is generated in the United States.

Preview Key Terms

In a notebook or on a separate sheet of paper, create a table like the one shown here to help with learning new key terms in the module. Before you read, fill out the "Prediction" column. Write what you think the term might mean or what it makes you think about. Use examples from your everyday life. There are no wrong answers!

Key Term	Prediction	Definition
Write key term here.	*Write what you think the term means in this column.*	*Define the term here. Add an example and use it in a sentence.*

Key Terms

Fossil fuel
Nonrenewable energy resource
Nuclear fuel
Commercial energy source
Subsistence energy source

Energy carrier
Turbine
Electrical grid
Combined cycle
Capacity

Capacity factor
Cogeneration
Combined heat and power

WHILE YOU READ THE MODULE

Define Key Terms

When you come across a new key term while reading the module, copy the definition into the "Definition" column of your key terms table. Add an example and use the term in a sentence. Compare your initial ideas to the actual definition.

Study the Figure

Examine Figure 34.1, "Worldwide annual energy consumption, by resource, in 2011" on page 400 and answer the following question.

1. What percentage of the worldwide annual energy consumption is nonrenewable energy?

Practice the Math: Efficiency of Travel

Read "Do the Math: Efficiency of Travel" on page 404. Try "Your Turn." For more math practice, do the following problem. Remember to show your work. Use a separate sheet of paper if needed.

If you could carpool with 2 other people from San Diego, California to Orlando, Florida, (a distance of roughly 3,910 k) what would the energy expenditure be per person? Would this be better than the other modes of transportation? Use the following information from the box on page 404.

Air   2.1 MJ/passenger-kilometer × 3,910 km/trip= 8,211 MJ/passenger trip
Car   3.6 MJ/passenger-kilometer × 3,910 km/trip= 14,076 MJ/passenger trip
Train   1.1 MJ/passenger-kilometer × 3,910 km/trip= 4,301 MJ/passenger trip
Bus   1.7 MJ/passenger-kilometer × 3,910 km/trip= 6,647 MJ/passenger trip

Study the Figure

Study Figure 34.7, "A coal-fired electricity generation plant" on page 406.

1. Without looking at the book, fill in the labels for 1-7 on the coal fired electricity generation plant depicted below. On a separate sheet of paper or in your notebook describe the role of each part in the generation and distribution of electricity.

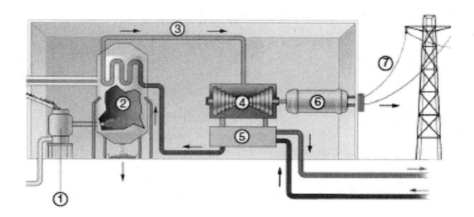

Practice the Math: Calculating Energy Supply

Read "Do the Math: Calculating Energy Supply," on page 407. Try "Your Turn." For more math practice do the following problem. Remember to show your work. Use a separate sheet of paper if needed.

According to the U.S. Department of Energy, a typical home in the United States uses approximately 900 kWh of electricity per month. On an annual basis, this is
900 kWh/month × 12 months/year = 10,800 kWh/year

During summer months in Alaska, some homes don't run an air conditioner very often. How many homes can the same power plant support if average electricity usage in Alaska decreases to 600 kWh/month during summer months?

AFTER YOU READ THE MODULE

Review Key Terms

Match the key terms on the left with the definitions on the right.

_____1. Fossil fuel

a. A power plant that uses both exhaust gases and steam turbines to generate electricity

_____2. Nonrenewable energy resource

b. Something that can move and deliver energy in a convenient, usable form to end users

_____3. Nuclear fuel

c. Fuel derived from radioactive materials that give off energy

_____4. Commercial energy source

d. An energy source gathered by individuals for their own immediate needs

_____5. Subsistence energy source

e. The use of a fuel to generate electricity and produce heat

_____6. Energy carrier

f. An energy source with a finite supply, primarily the fossil fuels and nuclear fuels

_____7. Turbine

g. The fraction of time a power plant operates in a year

_____8. Electrical grid

h. An energy source that is bought and sold

_____9. Combined cycle

i. In reference to an electricity-generating plant, the maximum electrical output

_____10. Capacity

j. A device with blades that can be turned by water, wind, steam, or exhaust gas from combustion that turns a generator in an electricity-producing plant

_____11. Capacity factor

k. A network of interconnected transmission lines that joins power plants together and links them with end users of electricity

_____12. Cogeneration (combined heat and power)

l. A fuel derived from biological material that became fossilized millions of years ago

Module 35: Fossil Fuel Resources

BEFORE YOU READ THE MODULE

Focus on Learning Objectives

Use the module learning objectives to guide your reading. On a separate piece of paper, write down each objective and take notes to help you meet each learning objective. After studying this module, you should be able to:

- discuss the uses of coal and its consequences.
- discuss the uses of oil and its consequences.
- discuss the uses of natural gas and its consequences.
- discuss the uses of oil sands and liquefied coal and its consequences.
- describe future prospects for fossil fuel use.

Preview Key Terms

In a notebook or on a separate sheet of paper, create a table like the one shown here to help with learning new key terms in the module. Before you read, fill out the "Prediction" column. Write what you think the term might mean or what it makes you think about. Use examples from your everyday life. There are no wrong answers!

Key Term	Prediction	Definition
Write key term here.	Write what you think the term means in this column.	Define the term here. Add an example and use it in a sentence.

Key Terms

Coal   Oil sands   Energy intensity
Petroleum   Bitumen   Hubbert curve
Crude oil   CTL (coal to liquid)   Peak oil

WHILE YOU READ THE MODULE

Define Key Terms

When you come across a new key term while reading the module, copy the definition into the "Definition" column of your key terms table. Add an example and use the term in a sentence. Compare your initial ideas to the actual definition.

Study The Figure

Examine Figure 35.7, "U.S. energy use per capita and energy intensity" on page 415 and answer the following questions.

1. What is the percent change in per capita energy use from 1950 to 2010?

2. What is the relationship from 1950 to 2010 between per capita energy use and energy use per dollar of GDP?

Review Key Terms

Match the key terms on the left with the definitions on the right.

_____1. Coal

_____2. Petroleum

_____3. Crude oil

_____4. Oil sands

_____5. Bitumen

_____6. CTL (coal to liquid)

_____7. Energy intensity

_____8. Hubbert curve

_____9. Peak oil

a. The process of converting solid coal into liquid fuel

b. Slow-flowing, viscous deposits of bitumen mixed with sand, water, and clay

c. The energy use per unit of gross domestic product

d. Liquid petroleum removed from the ground

e. A bell-shaped curve representing oil use and projecting both when world oil production will reach a maximum and when the world will run out of oil

f. The point at which half the total known oil supply is used up

g. A fossil fuel that occurs in underground deposits, composed of a liquid mixture of hydrocarbons, water, and sulfur

h. A degraded petroleum that forms when petroleum migrates to the surface of Earth and is modified by bacteria

i. A solid fuel formed primarily from the remains of trees, ferns, and other plant materials preserved 280 million to 360 million years ago

Module 36: Nuclear Energy Resources

BEFORE YOU READ THE MODULE

Focus on Learning Objectives

Use the module learning objectives to guide your reading. On a separate piece of paper, write down each objective and take notes to help you meet each learning objective. After studying this module, you should be able to:

- describe how nuclear energy is used to generate electricity.
- discuss the advantages and disadvantages of using nuclear fuels to generate electricity.

Preview Key Terms

In a notebook or on a separate sheet of paper, create a table like the one shown here to help with learning new key terms in the module. Before you read, fill out the "Prediction" column. Write what you think the term might mean or what it makes you think about. Use examples from your everyday life. There are no wrong answers!

Key Term	Prediction	Definition
Write key term here.	*Write what you think the term means in this column.*	*Define the term here. Add an example and use it in a sentence.*

Key Terms

Fission
Fuel rod
Control rod

Radioactive waste
Becquerel (Bq)
Curie

Nuclear fusion

WHILE YOU READ THE MODULE

Define Key Terms

When you come across a new key term while reading the module, copy the definition into the "Definition" column of your key terms table. Add an example and use the term in a sentence. Compare your initial ideas to the actual definition.

Practice the Math: Calculating Half Lives

Read "Do the Math: Calculating Half-Lives" on page 422. Try "Your Turn." For more math practice, do the following problems. Remember to show your work.

1. Uranium-235 has a half-live of 700 million years. How many years will it take Uranium-235 to decay to ⅛ of its original mass?

2. You have 150g of a radioactive substance. It has a half-life of 185 years. After 1,295 years, what mass remains?

AFTER YOU READ THE MODULE

Review Key Terms

Match the key terms on the left with the definitions on the right.

_____1. Fission

a. A cylindrical tube that encloses nuclear fuel within a nuclear reactor

_____2. Fuel rod

b. A unit of measure for radiation

_____3. Control rod

c. A nuclear reaction in which a neutron strikes a relatively large atomic nucleus, which then splits into two or more parts, releasing additional neutrons and energy in the form of heat

_____4. Radioactive waste

d. A reaction that occurs when lighter nuclei are forced together to produce heavier nuclei

_____5. Becquerel (Bq)

e. A cylindrical device inserted between the fuel rods in a nuclear reactor to absorb excess neutrons and slow or stop the fission reaction

_____6. Curie

f. Unit that measures the rate at which a sample of radioactive material decays

_____7. Nuclear fusion

g. Nuclear fuel that can no longer produce enough heat to be useful in a power plant but continues to emit radioactivity

Chapter ⑫ Review Exercises

Check Your Understanding

Review "Learning Objectives Revisited" on page 427 of your textbook. Compare the notes you took while reading each module. Complete these exercises to review the chapter.

1. What is the difference between commercial energy sources and subsistence energy sources?

2. What does Figure 34.3 tell you about energy consumption in the United States since 1850?

3. Explain why a bus is a more energy-efficient mode of travel than a car driven by a single passenger.

4. Summarize how a typical coal-fired electricity generation plant produces electricity.

5. Give an example of how a power plant could use cogeneration to obtain greater efficiencies.

6. Summarize the process of coal formation.

7. List 2 advantages and 2 disadvantages of using coal as an energy source.

8. List 2 advantages and 2 disadvantages of using petroleum as an energy source.

9. List 2 advantages and 2 disadvantages of using natural gas as an energy source.

10. List 2 advantages and 2 disadvantages of using nuclear power as an energy source.

Practice for Free-Response Questions

Complete this exercise to build and practice the skills you will need to answer free-response questions on the exam. Use a separate sheet of paper if necessary.

> One negative consequence of nuclear power is that it generates waste that must be stored. Deep underground storage sites have been suggested as storage locations for spent radioactive fuel. Identify and discuss two other options for storage of radioactive waste.

Review and Reflect

Complete these activities to solidify your knowledge of the chapter concepts and key terms. Use a notebook or a separate sheet of paper if necessary.

1. Review your key terms table for each module.

   (a) Which words did you already know? Which were new to you?
   (b) Write a new sentence using each key term.
   (c) Create a set of flash cards that includes each key term. Use the cards to review terms that were new or challenging.
   (d) When you feel comfortable with the new or challenging terms, review all of the cards, including those with familiar terms.
   (e) Save your cards to review before an exam.

2. What are the challenging concepts from this chapter?

   (a) Identify any concepts you found particularly challenging in this chapter.
   (b) Create a list of topics you need to review in preparation for an exam.

3. What questions do you have about concepts in the chapter?

   (a) Note any further questions you might have about material in the chapter.
   (b) Work with a partner to discuss these questions and ask your teacher for help as needed.

4. Write five possible multiple-choice questions based on this chapter. Work with a partner to quiz each other in preparation for an exam.

Chapter (13) Achieving Energy Sustainability

Chapter Summary

This chapter looks at renewable energy. Renewable energy is of great concern for our future use of energy as finite resources become depleted. The chapter also addresses the issue of energy conservation and efficiency. Major renewable energy sources considered include biomass, solar, hydroelectric, geothermal, wind, and fuel cell technology. As you go through the chapter, make sure you understand the advantages and disadvantages of each renewable energy source. The chapter concludes with a thoughtful consideration of our energy future. This is a great time for you to review both renewable and nonrenewable energy sources.
The chapter consists of 3 modules:

- **Module 37:** Conservation, Efficiency, and Renewable Energy
- **Module 38:** Biomass and Water
- **Module 39:** Solar, Wind, Geothermal, and Hydrogen
- **Module 40:** Planning Our Energy Future

Chapter Opening Case: *Energy from Wind*

The chapter opening case illustrates one of the most essential reasons to protect biodiversity: the use of resources for the development of new pharmaceutical drugs. Humans have already extracted life-saving drugs from a variety of species. With increasing rates of deforestation and habitat loss, many species that have never been researched for medical use are at risk of extinction. Furthermore, indigenous peoples with knowledge about medical uses of natural drugs are being forced to relocate. Their knowledge may soon be lost. This introductory case helps students understand the significance of biodiversity and the underlying mechanisms that allow organisms to adapt to their ever-changing environments.

Do the Math

This chapter contains the following "Do the Math" box to help prepare you for calculation questions you might encounter on the exam.

- "Energy Star" (page 436)

To make sure you understand the concepts and techniques presented in these boxes, do the practice problems presented in the text as well as the additional "Practice the Math" problems that appear in Module 37 of this study guide.

Module 37: Environmental Science

Focus on Learning Objectives

Use the module learning objectives to guide your reading. On a separate piece of paper, write down each objective and take notes to help you meet each learning objective. After studying this module, you should be able to:

- describe strategies to conserve energy and increase energy efficiency.
- explain differences among the various renewable energy resources.

Preview Key Terms

In a notebook or on a separate sheet of paper, create a table like the one shown here to help with learning new key terms in the module. Before you read, fill out the "Prediction" column. Write what you think the term might mean or what it makes you think about. Use examples from your everyday life. There are no wrong answers!

Key Term	Prediction	Definition
Write key term here.	*Write what you think the term means in this column.*	*Define the term here. Add an example and use it in a sentence.*

Key Terms

Biofuel	Carbon neutral	Biodiesel
Modern carbon	Net removal	Flex-fuel vehicle
Fossil carbon	Ethanol	Hydroelectricity

WHILE YOU READ THE MODULE

Define Key Terms

When you come across a new key term while reading the module, copy the definition into the "Definition" column of your key terms table. Add an example and use the term in a sentence. Compare your initial ideas to the actual definition.

Practice the Math: Energy Star

Read "Do the Math: Energy Star" on page 436. Try "Your Turn." For more math practice do the following problems. Remember to show your work.

1. You decide to purchase a new refrigerator. You have the choice of an Energy Star refrigerator for $2,000 or a standard unit for $1,800. The Energy Star unit costs $0.10 per hour less to run. If you buy the Energy Star unit and run it for 10 hours per day for a year, how long will it take you to recover the $200 extra cost?

2. You are about to purchase a new refrigerator. You can chose between an Energy Star model, which costs $300 and non-Energy Star model which costs $200. The cost of electricity is $0.10 per kilowatt-hour (kWh), and you expect your refrigerator to run an average of 10 hours per day.

    (a) The non-Energy Star model uses 0.5 kW. How much will it cost you per year for electricity to run this model?

    (b) If the Energy Star model uses 0.4 kW, how much money would you save on your electric bill over 5 years by buying the efficient model?

Study the Figure

Examine Figure 37.7, "Energy use in the United States" on page 439 and answer the following question.

1. What percent of the total energy use per year in the United States comes from wind energy? After studying this figure, describe a set of systems within systems in the area where you live or somewhere you have visited.

Review Key Terms

Match the key terms on the left with the definitions on the right.

_____1. Energy conservation	a. The greatest quantity of energy used at any one time.
_____2. Tiered rate system	b. Finding and implementing ways to use less energy
_____3. Peak demand	c. Construction designed to take advantage of solar radiation without active technology
_____4. Passive solar design	d. A billing system used by some electric companies in which customers pay higher rates as their use goes up
_____5. Thermal mass	e. A property of a building material that allows it to maintain heat or cold
_____6. Potentially renewable	f. An energy source that cannot be used up
_____7. Nondepletable	g. In energy management, an energy source that is either potentially renewable or nondepletable
_____8. Renewable	h. An energy source that can be regenerated indefinitely as long as it is not overharvested

Module 38: Biomass and Water

BEFORE YOU READ THE MODULE

Focus on Learning Objectives

Use the module learning objectives to guide your reading. On a separate piece of paper, write down each objective and take notes to help you meet each learning objective. After studying this module, you should be able to:

- describe the various forms of biomass.
- explain how energy is harnessed from water.

Preview Key Terms

In a notebook or on a separate sheet of paper, create a table like the one shown here to help with learning new key terms in the module. Before you read, fill out the "Prediction" column. Write what you think the term might mean or what it makes you think about. Use examples from your everyday life. There are no wrong answers!

Key Term	Prediction	Definition
Write key term here.	*Write what you think the term means in this column.*	*Define the term here. Add an example and use it in a sentence.*

Key Terms

Biofuel

Modern carbon

Fossil carbon

Carbon neutral

Net removal

Ethanol

Biodiesel

Flex-fuel vehicle

Hydroelectricity

Run-of-the-river

Water impoundment

Tidal energy

Siltation

WHILE YOU READ THE MODULE

Define Key Terms

When you come across a new key term while reading the module, copy the definition into the "Definition" column of your key terms table. Add an example and use the term in a sentence. Compare your initial ideas to the actual definition.

Study The Figure

Examine Figure 38.1, "Energy from the Sun" on page 441.

1. Which energy sources depicted in the figure do not come from the Sun?

Review Key Terms

Match the Key Terms on the left with the definitions on the right.

_____1. Biofuel

a. An activity that does not change atmospheric CO_2 concentrations

_____2. Modern carbon

b. A diesel substitute produced by extracting and chemically altering oil from plants

_____3. Fossil carbon

c. Liquid fuel created from processed or refined biomass

_____4. Carbon neutral

d. The process of removing more than is replaced by growth, typically used when referring to carbon

_____5. Net removal

e. Electricity generated by the kinetic energy of moving water

_____6. Ethanol

f. The storage of water in a reservoir behind a dam

_____7. Biodiesel

g. Carbon in biomass that was recently in the atmosphere

_____8. Flex-fuel vehicle

h. Alcohol made by converting starches and sugars from plant material into alcohol and CO_2

_____9. Hydroelectricity

i. Energy that comes from the movement of water driven by the gravitational pull of the Moon

_____10. Run-of-the-river

j. Carbon in fossil fuels

_____11. Water impoundment

k. A vehicle that runs on either gasoline or a gasoline/ethanol mixture

_____12. Tidal energy

l. Hydroelectricity generation in which water is retained behind a low dam or no dam

_____13. Siltation

m. The accumulation of sediments, primarily silt, on the bottom of a reservoir

Module 39: Solar, Wind, Geothermal, and Hydrogen

BEFORE YOU READ THE MODULE

Focus on Learning Objectives

Use the module learning objectives to guide your reading. On a separate piece of paper, write down each objective and take notes to help you meet each learning objective. After studying this module, you should be able to:

- list the different forms of solar energy and their application.
- describe how wind energy is harnessed and its contemporary uses.
- discuss the methods of harnessing the internal energy from Earth.
- explain the advantages and disadvantages of energy from hydrogen.

Preview Key Terms

In a notebook or on a separate sheet of paper, create a table like the one shown here to help with learning new key terms in the module. Before you read, fill out the "Prediction" column. Write what you think the term might mean or what it makes you think about. Use examples from your everyday life. There are no wrong answers!

Key Term	Prediction	Definition
Write key term here.	*Write what you think the term means in this column.*	*Define the term here. Add an example and use it in a sentence.*

Key Terms

Active solar energy
Photovoltaic solar cell
Wind energy
Wind turbine

Geothermal energy
Ground source heat pump
Fuel cell
Electrolysis

WHILE YOU READ THE MODULE

Define Key Terms

When you come across a new key term while reading the module, copy the definition into the "Definition" column of your key terms table. Add an example and use the term in a sentence. Compare your initial ideas to the actual definition.

Study the Figure

Examine Figure 39.7, "Installed wind energy capacity by country" on page 454 and answer the following questions.

1. Why is China the largest producer of wind energy but the holds the smallest percentage of electricity generated by wind in comparison to electricity generation in general for the country?

2. Portugal produced 5 GW of energy from the wind, which accounts for 20 percent of its total energy use. How much total energy does Portugal produce?

AFTER YOU READ THE MODULE

Review Key Terms

Match the key terms on the left with the definitions on the right.

_____1. Active solar energy

a. A turbine that converts wind energy into electricity

_____2. Photovoltaic solar cell

b. A system of capturing energy from sunlight and converting it directly into electricity

_____3. Wind energy

c. Energy generated from the kinetic energy of moving air

_____4. Wind turbine

d. A technology that transfers heat from the ground to a building

_____5. Geothermal energy

e. Energy captured from sunlight with advanced technologies

_____6. Ground source heat pump

f. The application of an electric current to water molecules to split them into hydrogen and oxygen

_____7. Fuel cell

g. Heat energy that comes from the natural radioactive decay of elements deep within Earth

_____8. Electrolysis

h. An electrical-chemical device that converts fuel, such as hydrogen, into an electrical current

Module 40: Planning Our Energy Future

BEFORE YOU READ THE MODULE

Focus on Learning Objectives

Use the module learning objectives to guide your reading. On a separate piece of paper, write down each objective and take notes to help you meet each learning objective. After studying this module, you should be able to:

- discuss the environmental and economic options we must assess in planning our energy future.
- consider the challenges of a renewable energy strategy.

Preview Key Terms

In a notebook or on a separate sheet of paper, create a table like the one shown here to help with learning new key terms in the module. Before you read, fill out the "Prediction" column. Write what you think the term might mean or what it makes you think about. Use examples from your everyday life. There are no wrong answers!

Key Term	Prediction	Definition
Write key term here.	*Write what you think the term means in this column.*	*Define the term here. Add an example and use it in a sentence.*

Key Term

Smart grid

WHILE YOU READ THE MODULE

Define Key Terms

When you come across a new key term while reading the module, copy the definition into the "Definition" column of your key terms table. Add an example and use the term in a sentence. Compare your initial ideas to the actual definition.

Chapter (13) Review Exercises

Check Your Understanding

Review "Learning Objectives Revisited" on page 467 of your textbook. Compare the notes you took while reading each module. Complete these exercises to review the chapter.

1. List three ways you could conserve energy.

2. What are some ways to utilize passive solar design?

3. Explain the difference between modern carbon and fossil carbon.

4. What are some items that can be turned into ethanol and biodiesel?

5. Fill in the following chart.

Energy Resource	Advantage	Disadvantage
Liquid biofuels		
Solid biomass		
Photovoltaic solar cells		
Solar water heating systems		
Hydroelectricity		
Tidal energy		
Geothermal energy		
Wind energy		
Hydrogen fuel cell		

Practice for Free-Response Questions

Complete this exercise to build and practice the skills you will need to answer free-response questions on the exam. Use a separate sheet of paper if necessary.

What problems do dams cause for fish? Name one solution and describe benefits and disadvantages.

Review and Reflect

Complete these activities to solidify your knowledge of the chapter concepts and key terms. Use a notebook or a separate sheet of paper if necessary.

1. Review your key terms table for each module.

   (a) Which words did you already know? Which were new to you?
   (b) Write a new sentence using each key term.
   (c) Create a set of flash cards that includes each key term. Use the cards to review terms that were new or challenging.
   (d) When you feel comfortable with the new or challenging terms, review all of the cards, including those with familiar terms.
   (e) Save your cards to review before an exam.

2. What are the challenging concepts from this chapter?

   (a) Identify any concepts you found particularly challenging in this chapter.
   (b) Create a list of topics you need to review in preparation for an exam.

3. What questions do you have about concepts in the chapter?

   (a) Note any further questions you might have about material in the chapter.
   (b) Work with a partner to discuss these questions and ask your teacher for help as needed.

4. Write five possible multiple-choice questions based on this chapter. Work with a partner to quiz each other in preparation for an exam.

Unit 6 Multiple-Choice Review Exam

Choose the best answer.

1. If the average person in the United States uses 10,000 watts of energy 24 hours a day for 365 days per year, how many kW of energy does the average person use in a year?
   (A) 10 kW
   (B) 1000 kW
   (C) 3,650 kW
   (D) 3,650,000 kW
   (E) 87,600 kW

2. Developed countries contain ____percent of the world's population and use ____ percent of the world's energy each year.
   (A) 5; 50
   (B) 20; 70
   (C) 5; 70
   (D) 1; 50
   (E) 50; 90

3. The energy source that is used most in the United States is
   (A) coal.
   (B) oil.
   (C) natural gas.
   (D) nuclear.
   (E) renewables.

4. Fuel efficiency (mpg) of U.S. automobiles has
   (A) decreased in the last 5 years.
   (B) increased in the past 30 years.
   (C) decreased and then increased in the past 30 years.
   (D) remained relatively stable since 1990.
   (E) dropped dramatically since 1990.

5. Which is NOT part of a coal-fired electricity generation plant?
   (A) Pulverizer
   (B) Boiler
   (C) Control rods
   (D) Turbine
   (E) Generator

6. Which is an example of cogeneration?
   (A) Using steam that is generated to heat buildings to turn a turbine
   (B) Using both coal and oil to create electricity
   (C) Increasing nuclear power plants in major metropolitan areas
   (D) Substituting anthracite coal for low grade lignite coal
   (E) Using both exhaust gases and steam turbines to generate electricity

7. Which lists types of coal in order from most moisture and least heat to least moisture and most heat?
   (A) Peat, lignite, bituminous, anthracite
   (B) Peat, bituminous, lignite, anthracite
   (C) Anthracite, bituminous, lignite, peat
   (D) Bituminous, anthracite, lignite, peat
   (E) Bituminous, lignite, peat, anthracite

8. Which of the following is correct?
      I. Petroleum comes from the remains of ocean-dwelling phytoplankton that died millions of years ago.
      II. Petroleum is found in porous, sedimentary rock.
      III. Petroleum burns cleaner than natural gas.
   (A) I only
   (B) II only
   (C) III only
   (D) I and II
   (E) I, II, and III

9. Natural gas is generally found with
   (A) oil.
   (B) coal.
   (C) both oil and coal.
   (D) uranium.
   (E) water.

10. Coal supplies are expected to last for at least
   (A) 5 years.
   (B) 40 years.
   (C) 60 years.
   (D) 100 years.
   (E) 200 years.

11. 1 gram of Uranium-235 contains _____ times the energy of 1 gram of coal.
   (A) 20-30
   (B) 1,000
   (C) 100,000-200,000
   (D) 2-3 million
   (E) 1 billion

12. Which is one difference between generating electricity with coal and generating electricity with nuclear energy?
   (A) Coal power generates steam and nuclear power does not.
   (B) Nuclear power uses fission to create heat to generate steam and coal does not.
   (C) Nuclear power produces more air pollution than coal.
   (D) Coal is much more energy efficient than nuclear energy.
   (E) A generator is not needed in the production of nuclear energy.

13. If a material has a radioactivity level of 100 curies and has a half-life of 10 years, how many half-lives will have occurred after 100 years?
    (A) 1 half-life
    (B) 4 half-lives
    (C) 10 half-lives
    (D) 1,000 half-lives
    (E) 25 half-lives

14. Which is a nonrenewable energy source?
    (A) Wind
    (B) Wood
    (C) Solar
    (D) Geothermal
    (E) Nuclear

15. Planting a large, deciduous shade tree next to a southern window is an example of
    (A) active solar design.
    (B) photovoltaic systems.
    (C) energy star technology.
    (D) passive solar design.
    (E) smart grid.

16. What major environmental impact is associated with deforestation?
    (A) Erosion
    (B) Acid rain
    (C) Depletion of the stratospheric ozone layer
    (D) Municipal solid waste
    (E) Vulnerability to invasive species

17. Most ethanol produced in the United States comes from _____ while in Brazil most ethanol comes from _____.
    (A) corn; oil deposits
    (B) oil sands; sugarcane
    (C) corn; sugarcane
    (D) biomass; wood chips
    (E) wood chips; corn

18. Which is an environmental consequence of a dam?
        I. Release of greenhouse gases
        II. Disruption to the life cycle of aquatic species
        III. Accumulation of sediments in the reservoir
    (A) I only
    (B) II only
    (C) III only
    (D) I and III only
    (E) II and III

19. A photovoltaic cell is used to
    (A) capture sunlight and turn it into electricity.
    (B) burn biomass fuel.
    (C) generate passive solar energy.
    (D) generate wind power.
    (E) generate electricity behind a dam.

20. Which countries lead the world in the production of geothermal energy?
    (A) The United States, Denmark, and Finland
    (B) Iceland, the United States, and China
    (C) China, Russia, and Denmark
    (D) Russia, The United States, and China
    (E) China, Finland, and Iceland

21. Which is NOT a benefit of wind power?
    (A) It requires no ongoing cost to harvest energy.
    (B) It does not harm wildlife.
    (C) It is relatively inexpensive.
    (D) It is renewable.
    (E) It can share land with other uses.

22. A hydrogen fuel cell uses hydrogen and oxygen to make
        I. electricity.
        II. energy.
        III. water.
    (A) I only
    (B) II only
    (C) III only
    (D) I and II
    (E) I, II, and III

23. A flex-fuel vehicle
    (A) is known as a hybrid.
    (B) uses hydrogen fuel cells for power.
    (C) runs on gasoline or E-85.
    (D) can drive farther on one tank of gas than a standard vehicle.
    (E) runs on photovoltaic cells.

24. Which is a disadvantage of tidal energy?
    (A) It is extremely expensive to run.
    (B) It generates carbon dioxide, a greenhouse gas.
    (C) It is aesthetically displeasing.
    (D) It is geographically limited.
    (E) It needs a storage battery to run.

25. Energy experts propose that our electrical infrastructure would be better if
    (A) we had larger, more centralized power generation plants.
    (B) we used more solar sources.
    (C) we switched to geothermal sources.
    (D) we had a larger number of small-scale power generation plants.
    (E) we switched to hydrogen fuel cells.

26. Which can NOT be substituted by biofuels such as ethanol and biodiesel?
    (A) Coal
    (B) Gasoline
    (C) Diesel
    (D) Petroleum
    (E) Oil

27. Which is an NOT environmental problem associated with the use of nuclear power?
    (A) High levels of air pollution during operation
    (B) Storage of spent fuel
    (C) The large amount of heat generated during operation
    (D) Mining to obtain nuclear fuel
    (E) Risk of nuclear accidents

28. Which is NOT an advantage of active solar technology?
    (A) It is a nondepletable resource.
    (B) After the initial investment there is no cost to harvest the energy.
    (C) It allows the owner to connect to the main grid to sell excess energy.
    (D) It is expensive to manufacture and install.
    (E) Energy produced by photovoltaic systems can be used in several ways.

29. A sample of radioactive waste has a half-life of 20 years and an activity level of 4 curies. How many years will it take for the activity level of this sample be 0.5 curies?
    (A) 40 years
    (B) 60 years
    (C) 80 years
    (D) 100 years
    (E) 120 years

Chapter (14) Water Pollution

Chapter Summary

This chapter introduces students to many different ways that water can be compromised. All organisms, including humans, require water to live, but increases in human population combined with industrialization have led to the contamination of water supplies. The chapter explores the main sources of water pollution: wastewater from humans and livestock, heavy metals from industry, oil pollution from mining and transportation, and nonchemical pollution such as thermal pollution and sediment. The chapter also describes ways in which water pollution can be remediated and discusses U.S. federal laws designed to protect waterways.
The chapter consists of 5 modules:

- **Module 41:** Wastewater from Humans and Livestock
- **Module 42:** Heavy Metals and Other Chemicals
- **Module 43:** Oil Pollution
- **Module 44:** Nonchemical Water Pollution
- **Module 45:** Water Pollution Laws

Chapter Opening Case: *The Chesapeake Bay*

The chapter opening case introduces you to how water pollution can affect aquatic ecosystems. The example of the Chesapeake Bay illustrates how nutrients, sediments, and chemicals in the water coming into the bay can affect different populations. It also demonstrates how several states and the federal government were able to work together and make substantial progress in cleaning up the bay.

Do the Math

This chapter contains the following "Do the Math" box to help prepare you for calculation questions you might encounter on the exam.

- "Building a Manure Lagoon" (page 489)

To make sure you understand the concepts and techniques presented in this box, do the practice problems presented in the text as well as the additional "Practice the Math" problems that appear in Module 41 of this study guide.

Module 41: Wastewater from Humans and Livestock

BEFORE YOU READ THE MODULE

Focus on Learning Objectives

Use the module learning objectives to guide your reading. On a separate piece of paper, write down each objective and take notes to help you meet each learning objective. After studying this module, you should be able to:

- discuss the three major problems caused by wastewater pollution.
- explain the modern technologies used to treat wastewater.

Preview Key Terms

In a notebook or on a separate sheet of paper, create a table like the one shown here to help with learning new key terms in the module. Before you read, fill out the "Prediction" column. Write what you think the term might mean or what it makes you think about. Use examples from your everyday life. There are no wrong answers!

Key Term	Prediction	Definition
Write key term here.	*Write what you think the term means in this column.*	*Define the term here. Add an example and use it in a sentence.*

Key Terms

Water pollution
Point source
Nonpoint source
Wastewater
Biochemical oxygen demand (BOD)
Dead zone
Eutrophication
Cultural eutrophication

Indicator species
Fecal coliform bacteria
Septic system
Septic tank
Sludge
Septage
Leach field
Manure lagoon

WHILE YOU READ THE MODULE

Define Key Terms

When you come across a new key term while reading the module, copy the definition into the "Definition" column of your key terms table. Add an example and use the term in a sentence. Compare your initial ideas to the actual definition.

Study the Figure

Examine Figure 41.6, "A sewage treatment plant" on page 488 and answer the following question.

1. How does a sewage treatment plant create a solid waste environmental problem?

Practice the Math: Building a Manure Lagoon

Read "Do the Math: Building a Manure Lagoon" on page 489. Try "Your Turn." For more math practice, do the following problems. Remember to show your work.

1. An animal produces 60 L of manure each day and the average number of animals on a feeding operation is 460 cattle.

    (a) How much manure is produced each day?

    (b) How much manure is produced in a week?

    (c) How much manure is produced in a year?

2. If an individual animal produces 50 L of manure each day and a manure lagoon needs to hold 45 days' worth of manure production for 1,500 cattle, what is the minimum capacity of the lagoon a farmer would need?

3. After the manure has broken down, the manure must be spread onto farm fields. A modern manure spreader can hold 40,000 L of liquid manure. How many trips will it take for the manure spreader to remove the 45 days' worth of manure that is held in the manure lagoon?

AFTER YOU READ THE MODULE

Review Key Terms

Match the key terms on the left with the definitions on the right.

_____1. Water pollution

_____2. Wastewater

_____3. Point source

_____4. Nonpoint source

_____5. Biochemical oxygen demand (BOD)

_____6. Dead zone

_____7. Eutrophication

_____8. Cultural eutrophication

_____9. Indicator species

_____10. Fecal coliform bacteria

_____11. Septic system

a. A phenomenon in which a body of water becomes rich in nutrients

b. A diffuse area that produces pollution

c. An increase in fertility in a body of water, the result of anthropogenic inputs of nutrients

d. The amount of oxygen a quantity of water uses over a period of time at specific temperatures

e. A species that indicates whether or not disease-causing pathogens are likely to be present

f. In a body of water, an area with extremely low oxygen concentration and very little life

g. A group of generally harmless microorganisms in human intestines that can serve as an indicator species for potentially harmful microorganisms associated with contaminated sewage

h. A relatively small and simple sewage treatment system, made up of a septic tank and a leach field, often used for homes in rural areas

i. A large container that receives wastewater from a house as part of a septic system

j. The contamination of streams, rivers, lakes, oceans, or groundwater with substances produced through human activities

k. A component of a septic system, made up of underground pipes laid out below the surface of the ground

_____12. Septic tank

l. A distinct location from which pollution is directly produced

_____13. Sludge

m. Solid waste material from wastewater

_____14. Septage

n. Water produced by livestock operations and human activities, including human sewage from toilets and gray water from bathing and washing of clothes and dishes

_____15. Leach field

o. A layer of fairly clear water found in the middle of a septic tank

_____16. Manure lagoon

p. Human-made pond lined with rubber built to handle large quantities of manure produced by livestock

Module 42: Heavy Metals and Other Chemicals

BEFORE YOU READ THE MODULE

Focus on Learning Objectives

Use the module learning objectives to guide your reading. On a separate piece of paper, write down each objective and take notes to help you meet each learning objective. After studying this module, you should be able to:

- identify key environmental indicators and their trends over time.
- define sustainability and explain how it can be measured using the ecological footprint.

Preview Key Terms

In a notebook or on a separate sheet of paper, create a table like the one shown here to help with learning new key terms in the module. Before you read, fill out the "Prediction" column. Write what you think the term might mean or what it makes you think about. Use examples from your everyday life. There are no wrong answers!

Key Term	Prediction	Definition
Write key term here.	_Write what you think the term means in this column._	_Define the term here. Add an example and use it in a sentence._

Key Terms

Acid deposition
Perchlorates
Polychlorinated biphenyls (PCBs)

Define Key Terms

When you come across a new key term while reading the module, copy the definition into the "Definition" column of your key terms table. Add an example and use the term in a sentence. Compare your initial ideas to the actual definition.

AFTER YOU READ THE MODULE

Review Key Terms

Match the key terms on the left with the definitions on the right.

_____1. Acid deposition   a. A group of harmful chemicals used for rocket fuel

_____2. Perchlorates   b. Acids deposited on Earth as rain and snow or as gases and particles that attach to the surfaces of plants, soil, and water

_____3. Polychlorinated biphenyls (PCBs)   c. A group of industrial compounds used to manufacture plastics and insulate electrical transformers, and responsible for many environmental problems

Module 43: Oil Pollution

BEFORE YOU READ THE MODULE

Focus on Learning Objectives

Use the module learning objectives to guide your reading. On a separate piece of paper, write down each objective and take notes to help you meet each learning objective. After studying this module, you should be able to:

- identify the major sources of oil pollution.
- explain some of the current methods to remediate oil pollution.

WHILE YOU READ THE MODULE

Study the Figure

Examine Figure 43.2, "Sources of oil in the ocean" on page 499. Use a separate sheet of paper if needed.

1. Why is there less oil pollution from the extraction of petroleum in North America compared to the world?

Module 44: Nonchemical Water Pollution

BEFORE YOU READ THE MODULE

Focus on Learning Objectives

Use the module learning objectives to guide your reading. On a separate piece of paper, write down each objective and take notes to help you meet each learning objective. After studying this module, you should be able to:

- identify the major sources of solid waste pollution.
- explain the harmful effect of sediment pollution.
- discuss the sources and consequences of thermal pollution.
- understand the causes of noise pollution.

Preview Key Terms

In a notebook or on a separate sheet of paper, create a table like the one shown here to help with learning new key terms in the module. Before you read, fill out the "Prediction" column. Write what you think the term might mean or what it makes you think about. Use examples from your everyday life. There are no wrong answers!

Key Term	Prediction	Definition
Write key term here.	*Write what you think the term means in this column.*	*Define the term here. Add an example and use it in a sentence.*

Key Terms

Thermal pollution
Thermal shock

WHILE YOU READ THE MODULE

Define Key Terms

When you come across a new key term while reading the module, copy the definition into the "Definition" column of your key terms table. Add an example and use the term in a sentence. Compare your initial ideas to the actual definition.

Review Key Terms

Match the key terms on the left with the definitions on the right.

_____1. Thermal pollution

a. A dramatic change in water temperature that can kill organisms

_____2. Thermal shock

b. Nonchemical water pollution that occurs when human activities cause a substantial change in the temperature of water

Module 45: Water Pollution Laws

BEFORE YOU READ THE MODULE

Focus on Learning Objectives

Use the module learning objectives to guide your reading. On a separate piece of paper, write down each objective and take notes to help you meet each learning objective. After studying this module, you should be able to:

* explain how the Clean Water Act protects against water pollution.
* discuss the goals of the Safe Drinking Water Act.
* understand how water pollution legislation is changing in developing countries.

Preview Key Terms

In a notebook or on a separate sheet of paper, create a table like the one shown here to help with learning new key terms in the module. Before you read, fill out the "Prediction" column. Write what you think the term might mean or what it makes you think about. Use examples from your everyday life. There are no wrong answers!

Key Term	Prediction	Definition
Write key term here.	*Write what you think the term means in this column.*	*Define the term here. Add an example and use it in a sentence.*

Key Terms

Clean Water Act
Safe Drinking Water Act
Maximum contaminant level (MCL)

Define Key Terms

When you come across a new key term while reading the module, copy the definition into the "Definition" column of your key terms table. Add an example and use the term in a sentence. Compare your initial ideas to the actual definition.

AFTER YOU READ THE MODULE

Review Key Terms

Match the key terms on the left with the definitions on the right.

_____1. Clean Water Act

a. Legislation that sets the national standards for safe drinking water

_____2. Safe Drinking Water Act

b. The standard for safe drinking water established by the EPA under the Safe Drinking Water Act

_____3. Maximum contaminant level (MCL)

c. Legislation that supports the "protection and propagation of fish, shellfish, and wildlife and recreation in and on the water" by maintaining and, when necessary, restoring the chemical, physical, and biological properties of surface waters

Chapter ⑭ Review Exercises

Check Your Understanding

Review "Learning Objectives Revisited" on page 512 of your textbook. Compare the notes you took while reading each module. Complete these exercises to review the chapter.

1. Define point and nonpoint source pollutants and give examples of each.

2. What are the three reasons that environmental scientists are concerned about human wastewater as a pollutant?

3. Summarize the steps in a sewage treatment plant.

4. What are some sources of lead, arsenic, and mercury?

5. Explain how DDT worked its way up the food chain and give an example of how it affected non-pest species.

6. How can thermal pollution harm aquatic species?

Practice for Free-Response Questions

Complete this exercise to build and practice the skills you will need to answer free-response questions on the exam. Use a separate sheet of paper if necessary.

Identify and describe three approaches to remediate oil pollution.

Review and Reflect

Complete these activities to solidify your knowledge of the chapter concepts and key terms. Use a notebook or a separate sheet of paper if necessary.

1. Review your key terms table for each module.

    (a) Which words did you already know? Which were new to you?
    (b) Write a new sentence using each key term.
    (c) Create a set of flash cards that includes each key term. Use the cards to review terms that were new or challenging.
    (d) When you feel comfortable with the new or challenging terms, review all of the cards, including those with familiar terms.
    (e) Save your cards to review before an exam.

2. What are the challenging concepts from this chapter?

    (a) Identify any concepts you found particularly challenging in this chapter.
    (b) Create a list of topics you need to review in preparation for an exam.

3. What questions do you have about concepts in the chapter?

    (a) Note any further questions you might have about material in the chapter.
    (b) Work with a partner to discuss these questions and ask your teacher for help as needed.

4. Write five possible multiple-choice questions based on this chapter. Work with a partner to quiz each other in preparation for an exam.

Chapter (15) Air Pollution and Stratospheric Ozone Depletion

Chapter Summary

This chapter discusses air pollution, one of the most significant topics in the AP® environmental science course and a subject that you may find relevant to your life. Air pollution is a problem that occurs in both developing and developed countries. Because air is a common resource across Earth, air pollution crosses many system boundaries. Human activity contributes to both outdoor and indoor air pollution. This chapter identifies the major air pollutants found around the globe and examines the specific air pollution situations that occur with photochemical smog and acid deposition. The chapter looks at a variety of air pollution control measures and examines stratospheric ozone depletion, and concludes with a discussion of indoor air pollution. The chapter consists of 5 modules:

- **Module 46:** Major Air Pollutants and Their Sources
- **Module 47:** Photochemical Smog and Acid Rain
- **Module 48:** Pollution Control Measures
- **Module 49:** Stratospheric Ozone Depletion
- **Module 50:** Indoor Air Pollution

Chapter Opening Case: *Cleaning Up in Chattanooga*

The chapter opening discusses how the city of Chattanooga, Tennessee, has improved its air quality. One part of this case discusses how ground-level ozone concentrations came to be extremely high in the late 1990s. It goes on to discuss how the private and public sectors had to work together to help Chattanooga attain a lower ozone level. This case illustrates some of the challenges cities face in cleaning up pollution and the importance of cooperation between government and industry.

Do the Math

This chapter contains the following "Do the Math" box to help prepare you for calculation questions you might encounter on the exam.

- "Calculating Annual Sulfur Reductions" (page 536)

To make sure you understand the concepts and techniques presented in these boxes, do the practice problems presented in the text as well as the additional "Practice the Math" problems that appear in Module 48 of this study guide.

Module 46: Major Air Pollutants and Their Sources

BEFORE YOU READ THE MODULE

Focus on Learning Objectives

Use the module learning objectives to guide your reading. On a separate piece of paper, write down each objective and take notes to help you meet each learning objective. After studying this module, you should be able to:

- identify and describe the major air pollutants.
- describe the sources of air pollution.

Preview Key Terms

In a notebook or on a separate sheet of paper, create a table like the one shown here to help with learning new key terms in the module. Before you read, fill out the "Prediction" column. Write what you think the term might mean or what it makes you think about. Use examples from your everyday life. There are no wrong answers!

Key Term	Prediction	Definition
Write key term here.	*Write what you think the term means in this column.*	*Define the term here. Add an example and use it in a sentence.*

Key Terms

Air pollution
Particulate matter (PM)
Particulates
Particles
Haze
Photochemical oxidant
Ozone (O_3)
Smog
Photochemical smog

Los Angeles-type smog
Brown smog
Sulfurous smog
London-type smog
Gray smog
Industrial smog
Volatile organic compound (VOC)
Primary pollutant
Secondary pollutant

WHILE YOU READ THE MODULE

Define Key Terms

When you come across a new key term while reading the module, copy the definition into the "Definition" column of your key terms table. Add an example and use the term in a sentence. Compare your initial ideas to the actual definition.

Study the Figure

Use Figure 46.5, " Emission sources of criteria air pollutants for the United States" on page 525 to answer the questions below.

1. Of all the air pollutants produced by on-road vehicles, which is produced in the highest quantity?

2. Electricity generation is the smallest emission source for air pollutants in the United States. However, electricity generation produces more nitrogen oxides than any other source. Why does electricity generation produce such high volumes of nitrogen oxides?

AFTER YOU READ THE MODULE

Review Key Terms

Match the key terms on the left with the definitions on the right.

_____1. Air pollution

_____2. Particulate matter (PM)

_____3. Haze

_____4. Photochemical oxidant

_____5. Ozone (O$_3$)

_____6. Smog

_____7. Photochemical smog

_____8. Sulfurous smog

_____9. Volatile organic compound (VOC)

a. A secondary pollutant made up of three oxygen atoms bound together

b. An organic compound that evaporates at typical atmospheric temperatures

c. Smog that is dominated by oxidants such as ozone

d. The introduction of chemicals, particulate matter, or microorganisms into the atmosphere at concentrations high enough to harm plants, animals, and materials such as buildings, or to alter ecosystems

e. Solid or liquid particles suspended in air

f. A primary pollutant that has undergone transformation in the presence of sunlight, water, oxygen, or other compounds

g. A polluting compound that comes directly out of a smokestack, exhaust pipe, or natural emission source

h. Reduced visibility

i. A type of air pollution that is a mixture of oxidants and particulate matter

_____10. Primary pollutant

j. Smog dominated by sulfur dioxide and sulfate compounds

_____11. Secondary pollutant

k. A class of air pollutants formed as a result of sunlight acting on compounds such as nitrogen oxides

Module 47: Photochemical Smog and Acid Rain

BEFORE YOU READ THE MODULE

Focus on Learning Objectives

Use the module learning objectives to guide your reading. On a separate piece of paper, write down each objective and take notes to help you meet each learning objective. After studying this module, you should be able to:

- explain how photochemical smog forms and why it is still a problem in the United States.
- describe how acid deposition forms and why it has improved in the United States and become worse elsewhere.

Preview Key Terms

In a notebook or on a separate sheet of paper, create a table like the one shown here to help with learning new key terms in the module. Before you read, fill out the "Prediction" column. Write what you think the term might mean or what it makes you think about. Use examples from your everyday life. There are no wrong answers!

Key Term	Prediction	Definition
Write key term here.	_Write what you think the term means in this column._	_Define the term here. Add an example and use it in a sentence._

Key Terms

Thermal inversion
Inversion layer

WHILE YOU READ THE MODULE

Define Key Terms

When you come across a new key term while reading the module, copy the definition into the "Definition" column of your key terms table. Add an example and use the term in a sentence. Compare your initial ideas to the actual definition.

Study The Figure

Examine Figure 47.1, "Tropospheric ozone and photochemical smog formation" on page 528. Answer the following questions.

1. Label the process of natural ozone accumulation, natural ozone destruction and build-up of photochemical smog in the figure below.

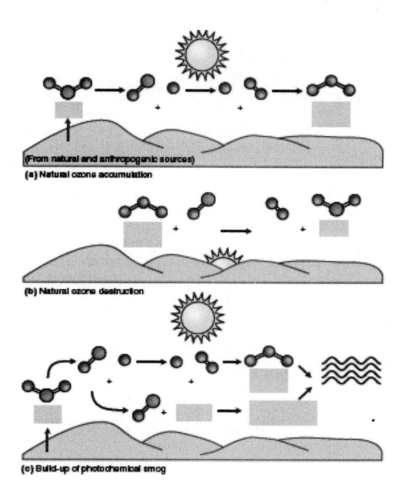

2. Name some everyday substances that produce VOCs.

Study The Figure

Study Figure 47.3, "Formation of acid deposition" on page 531. Answer the following questions.

1. Label the figure below with the appropriate primary and secondary pollutants.

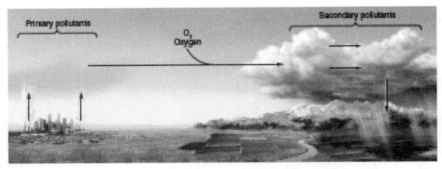

2. Describe some of the environmental impacts of acid deposition.

AFTER YOU READ THE MODULE

Review Key Terms

Match the Key Terms on the left with the definitions on the right.

_____1. Thermal inversion

a. The layer of warm air that traps emissions in a thermal inversion

_____2. Inversion layer

b. A situation in which a relatively warm layer of air at mid-altitude covers a layer of cold, dense air below

Module 48: Pollution Control Measures

BEFORE YOU READ THE MODULE

Focus on Learning Objectives

Use the module learning objectives to guide your reading. On a separate piece of paper, write down each objective and take notes to help you meet each learning objective. After studying this module, you should be able to:

- explain strategies and techniques for controlling sulfur dioxide, nitrogen oxides, and particulate matter.
- describe innovative pollution control measures.

WHILE YOU READ THE MODULE

Practice the Math: Calculating Annual Sulfur Reductions

Read "Do the Math: Calculating Annual Sulfur Reductions" on page 536. Try "Your Turn." For more math practice, do the following problems. Remember to show your work.

1. A researcher just published an article stating that CO_2 levels have increased by 8 ppm. If the level increased from 376 ppm to 384 ppm in 5 years, find the percent change of these emissions.

2. In the United States, carbon dioxide emissions increased from 19.2 metric tons in 1990 to 19.72 metric tons in 2005. Calculate the total percentage increase and the annual percentage reduction for carbon dioxide.

Module 49: Stratospheric Ozone Depletion

BEFORE YOU READ THE MODULE

Focus on Learning Objectives

Use the module learning objectives to guide your reading. On a separate piece of paper, write down each objective and take notes to help you meet each learning objective. After studying this module, you should be able to:

- explain the benefits of stratospheric ozone and how it forms.
- describe the depletion of stratospheric ozone.
- explain efforts to reduce ozone depletion.

Module 50: Indoor Air Pollution

BEFORE YOU READ THE MODULE

Focus on Learning Objectives

Use the module learning objectives to guide your reading. On a separate piece of paper, write down each objective and take notes to help you meet each learning objective. After studying this module, you should be able to:

- explain how photochemical smog forms and why it is still a problem in the United States.
- describe how acid deposition forms and why it has improved in the United States and become worse elsewhere.

Preview Key Terms

In a notebook or on a separate sheet of paper, create a table like the one shown here to help with learning new key terms in the module. Before you read, fill out the "Prediction" column. Write what you think the term might mean or what it makes you think about. Use examples from your everyday life. There are no wrong answers!

Key Term	Prediction	Definition
Write key term here.	*Write what you think the term means in this column.*	*Define the term here. Add an example and use it in a sentence.*

Key Terms

Asbestos
Sick building syndrome

WHILE YOU READ THE MODULE

Define Key Terms

When you come across a new key term while reading the module, copy the definition into the "Definition" column of your key terms table. Add an example and use the term in a sentence. Compare your initial ideas to the actual definition.

AFTER YOU READ THE MODULE

Review Key Terms

Match the Key Terms on the left with the definitions on the right.

_____1. Thermal inversion

    a. The layer of warm air that traps emissions in a thermal inversion

_____2. Inversion layer

    b. A situation in which a relatively warm layer of air at mid-altitude covers a layer of cold, dense air below

Chapter ⑮ Review Exercises

Check Your Understanding

Review "Learning Objectives Revisited" on page 548 of your textbook. Compare the notes you took while reading each module. Complete these exercises to review the chapter.

1. Fill in the following chart:

Compound	Effects
Sulfur dioxide	
Nitrogen dioxide	
Carbon monoxide	
Particulate matter	
Lead	
Ozone	
VOCs	
Mercury	
Carbon dioxide	

2. Explain the difference between photochemical smog and sulfurous smog.

3. What are some human causes of air pollution?

4. Explain the process of thermal inversion and its impact on air quality.

5. What are the effects of acid deposition?

6. What are some strategies that countries and cities have taken to lower car emissions?

7. Summarize the chemical reactions that destroy stratospheric ozone.

8. What were sources of CFCs? Why were they banned?

9. Name some indoor air pollutants.

Practice for Free-Response Questions

Complete this exercise to build and practice the skills you will need to answer free-response questions on the exam. Use a separate sheet of paper if necessary.

Describe both primary and secondary pollutants and provide an example of each.

Review and Reflect

Complete these activities to solidify your knowledge of the chapter concepts and key terms. Use a notebook or a separate sheet of paper if necessary.

1. Review your key terms table for each module.

    (a) Which words did you already know? Which were new to you?
    (b) Write a new sentence using each key term.
    (c) Create a set of flash cards that includes each key term. Use the cards to review terms that were new or challenging.
    (d) When you feel comfortable with the new or challenging terms, review all of the cards, including those with familiar terms.
    (e) Save your cards to review before an exam.

2. What are the challenging concepts from this chapter?

    (a) Identify any concepts you found particularly challenging in this chapter.
    (b) Create a list of topics you need to review in preparation for an exam.

3. What questions do you have about concepts in the chapter?

    (a) Note any further questions you might have about material in the chapter.
    (b) Work with a partner to discuss these questions and ask your teacher for help as needed.

4. Write five possible multiple-choice questions based on this chapter. Work with a partner to quiz each other in preparation for an exam.

Chapter (16) Waste Generation and Waste Disposal

Chapter Summary

Humans are the only organisms to generate waste that other organisms cannot use. The chapter begins by looking at how humans generate waste. Total municipal solid waste (MSW) and per capita waste generation in the United States have recently started to decrease. The chapter moves from how humans generate waste to ways in which we can divert material from the solid waste stream: reduction, reuse, recycling, and composting. The chapter also looks closely at the environmental costs and benefits of landfills and incineration. Hazardous waste is a special category of waste that must be handled and disposed of with particular care. National legislation that addresses hazardous waste includes the U.S. Resource Conservation and Recovery Act (RCRA) and the Comprehensive Environmental Response, Compensation, and Liability Act (CERCLA). The chapter continues with a detailed discussion of life-cycle analysis, which uses a holistic approach to study the entire waste stream from the creation of materials through their use and ultimate disposal. Finally, the chapter explores integrated waste management, which offers multiple approaches for reducing the waste stream. The chapter consists of 5 modules:

- **Module 51:** Only Humans Generate Waste
- **Module 52:** The Three Rs and Composting
- **Module 53:** Landfills and Incineration
- **Module 54:** Hazardous Waste
- **Module 55:** New Ways to Think About Solid Waste

Chapter Opening Case: *Paper or Plastic?*

The chapter opening case introduces you to the costs and benefits of using hydraulic fracturing, known as fracking, to extract oil and gas. This case demonstrates how human activities that are initially perceived as causing little harm to the environment can in fact have substantial adverse effects. It also illustrates the controversial side of issues that environmental scientists explore and the difficulty in obtaining absolute answers to environmental problems and questions.

Do the Math

This chapter contains the following "Do the Math" box to help prepare you for calculation questions you might encounter on the exam.

- "How Much Leachate Might Be Collected" (page 572)

To make sure you understand the concepts and techniques presented in these boxes, do the practice problems presented in the text as well as the additional "Practice the Math" problems that appear in Module 53 of this study guide.

Module 51: Environmental Science

BEFORE YOU READ THE MODULE

Focus on Learning Objectives

Use the module learning objectives to guide your reading. On a separate piece of paper, write down each objective and take notes to help you meet each learning objective. After studying this module, you should be able to:

- explain why we generate waste and describe recent waste disposal trends.
- describe the content of the solid waste stream in the United States.

Preview Key Terms

In a notebook or on a separate sheet of paper, create a table like the one shown here to help with learning new key terms in the module. Before you read, fill out the "Prediction" column. Write what you think the term might mean or what it makes you think about. Use examples from your everyday life. There are no wrong answers!

Key Term	Prediction	Definition
Write key term here.	*Write what you think the term means in this column.*	*Define the term here. Add an example and use it in a sentence.*

Key Terms

Waste
Municipal solid waste (MSW)
Waste stream

WHILE YOU READ THE MODULE

Define Key Terms

When you come across a new key term while reading the module, copy the definition into the "Definition" column of your key terms table. Add an example and use the term in a sentence. Compare your initial ideas to the actual definition.

Study the Figure

Use Figure 51.5, "Composition and sources of municipal solid waste (MSW) in the United States" on page 558 to answer the following questions.

1. What percentage of the total original waste stream is discarded?

2. What could be some reasons that only a fraction of compostable MSW is actually composted?

AFTER YOU READ THE MODULE

Review Key Terms

Match the key terms on the left with the definitions on the right.

_____1. Waste

a. The flow of solid waste that is recycled, incinerated, placed in a solid waste landfill, or disposed of in another way

_____2. Municipal solid waste (MSW)

b. Material outputs from a system that are not useful or consumed

_____3. Waste stream

c. Refuse collected by municipalities from households, small businesses, and institutions

Module 52: The Three Rs and Composting

BEFORE YOU READ THE MODULE

Focus on Learning Objectives

Use the module learning objectives to guide your reading. On a separate piece of paper, write down each objective and take notes to help you meet each learning objective. After studying this module, you should be able to:

- explain why we generate waste and describe recent waste disposal trends.
- describe the content of the solid waste stream in the United States.

Preview Key Terms

In a notebook or on a separate sheet of paper, create a table like the one shown here to help with learning new key terms in the module. Before you read, fill out the "Prediction" column. Write what you think the term might mean or what it makes you think about. Use examples from your everyday life. There are no wrong answers!

Key Term	Prediction	Definition
Write key term here.	*Write what you think the term means in this column.*	*Define the term here. Add an example and use it in a sentence.*

Key Terms

Reduce, reuse, recycle
The three Rs
Source reduction

Reuse
Recycling
Closed-loop recycling

Open-loop recycling
Compost

WHILE YOU READ THE MODULE

Define Key Terms

When you come across a new key term while reading the module, copy the definition into the "Definition" column of your key terms table. Add an example and use the term in a sentence. Compare your initial ideas to the actual definition.

Review Key Terms

Match the Key Terms on the left with the definitions on the right.

_____1. Reduce, reuse, recycle (the three Rs)

_____2. Source reduction

_____3. Reuse

_____4. Recycling

_____5. Closed-loop recycling

_____6. Open-loop recycling

_____7. Composting

a. Using a product or material that was intended to be discarded

b. A popular phrase promoting the idea of diverting materials from the waste stream

c. Recycling a product into the same product

d. Creation of organic matter (humus) by decomposition under controlled conditions to produce an organic-rich material that enhances soil structure, cation exchange capacity, and fertility

e. An approach to waste management that seeks to cut waste by reducing the use of potential waste materials in the early stages of design and manufacture

f. Recycling one product into a different product

g. The process by which materials destined to become municipal solid waste (MSW) are collected and converted into raw material that is then used to produce new objects

Module 53: Landfills and Incineration

BEFORE YOU READ THE MODULE

Focus on Learning Objectives

Use the module learning objectives to guide your reading. On a separate piece of paper, write down each objective and take notes to help you meet each learning objective. After studying this module, you should be able to:

- describe the goals and function of a solid waste landfill.
- explain the design and purpose of a solid waste incinerator.

Preview Key Terms

In a notebook or on a separate sheet of paper, create a table like the one shown here to help with learning new key terms in the module. Before you read, fill out the "Prediction" column. Write what you think the term might mean or what it makes you think about. Use examples from your everyday life. There are no wrong answers!

Key Term	Prediction	Definition
Write key term here.	*Write what you think the term means in this column.*	*Define the term here. Add an example and use it in a sentence.*

Key Terms

Leachate

Sanitary landfill

Tipping fee

Siting

Incineration

Ash

Bottom ash

Fly ash

Waste-to-energy

WHILE YOU READ THE MODULE

Define Key Terms

When you come across a new key term while reading the module, copy the definition into the "Definition" column of your key terms table. Add an example and use the term in a sentence. Compare your initial ideas to the actual definition.

Study the Figure

Study Figure 53.2, "A modern sanitary landfill" on page 569. Answer the following questions.

1. Label the parts of a modern sanitary landfill in the figure below.

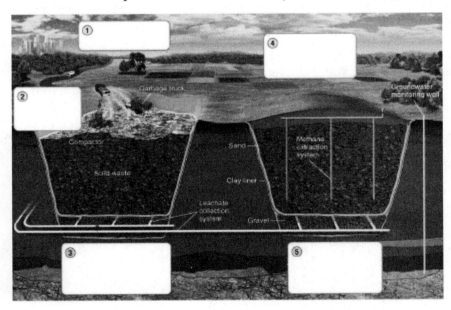

2. Identify three systems that are put into place in a sanitary landfill to stop leachate from contaminating groundwater.

Do the Math: How Much Leachate Might Be Collected

Read "Do the Math: How Much Leachate Might Be Collected," on page 572. Try "Your Turn." For more math practice, do the following problems. Use a separate sheet of paper if needed.

1. The annual precipitation at a landfill is 250mm per year, and 50 percent of this water runs off the landfill.

    (a) If the landfill has a surface area of 10,000 m^2 and the leachate collection system is 80 percent effective, calculate the volume of water in cubic meters that infiltrates the landfill per year.

    (b) How much volume of leachate in m^3 is treated per year?

2. Annual precipitation at a landfill is 100 mm per year, and 60 percent of this water runs off the landfill without infiltrating the surface. The landfill has a surface area of 5,000 m^2. The town has installed a leachate collection system beneath the landfill. If the collection system is 70 percent effective, what is the volume of water in cubic meters (m^3) that infiltrates the landfill per year?

Review Key Terms

Match the key terms on the left with the definitions on the right.

_____1. Leachate

_____2. Sanitary landfill

_____3. Tipping fee

_____4. Siting

_____5. Incineration

_____6. Ash

_____7. Bottom ash

_____8. Fly ash

_____9. Waste-to-energy

a. The residual nonorganic material that does not combust during incineration

b. A system in which heat generated by incineration is used as an energy source rather than released into the atmosphere

c. The residue collected from the chimney or exhaust pipe of a furnace

d. An engineered ground facility designed to hold municipal solid waste (MSW) with as little contamination of the surrounding environment as possible

e. The designation of a landfill location, typically through a regulatory process involving studies, written reports, and public hearings

f. Liquid that contains elevated levels of pollutants as a result of having passed through municipal solid waste (MSW) or contaminated soil

g. A fee charged for disposing of material in a landfill or incinerator

h. The process of burning waste materials to reduce volume and mass, sometimes to generate electricity or heat

i. Residue collected at the bottom of the combustion chamber in a furnace

Module 54: Hazardous Waste

BEFORE YOU READ THE MODULE

Focus on Learning Objectives

Use the module learning objectives to guide your reading. On a separate piece of paper, write down each objective and take notes to help you meet each learning objective. After studying this module, you should be able to:

- define hazardous waste and discuss the issues involved in handling it.
- describe regulations and legislation regarding hazardous waste.

Preview Key Terms

In a notebook or on a separate sheet of paper, create a table like the one shown here to help with learning new key terms in the module. Before you read, fill out the "Prediction" column. Write what you think the term might mean or what it makes you think about. Use examples from your everyday life. There are no wrong answers!

Key Term	Prediction	Definition
Write key term here.	*Write what you think the term means in this column.*	*Define the term here. Add an example and use it in a sentence.*

Key Terms

Hazardous waste
Superfund Act
Brownfields

WHILE YOU READ THE MODULE

Define Key Terms

When you come across a new key term while reading the module, copy the definition into the "Definition" column of your key terms table. Add an example and use the term in a sentence. Compare your initial ideas to the actual definition.

AFTER YOU READ THE MODULE

Review Key Terms

Match the key terms on the left with the definitions on the right.

_____1. Hazardous waste

a. The common name for the Comprehensive Environmental Response, Compensation, and Liability Act (CERCLA); a 1980 U.S. federal act that imposes a tax on the chemical and petroleum industries, funds the cleanup of abandoned and nonoperating hazardous waste sites, and authorizes the federal government to respond directly to the release or threatened release of substances that may pose a threat to human health or the environment

_____2. Superfund Act

b. Contaminated industrial or commercial sites that may require environmental cleanup before they can be redeveloped or expanded

_____3. Brownfields

c. Liquid, solid, gaseous, or sludge waste material that is harmful to humans or ecosystems

Module 55: New Ways to Think About Solid Waste

BEFORE YOU READ THE MODULE

Focus on Learning Objectives

Use the module learning objectives to guide your reading. On a separate piece of paper, write down each objective and take notes to help you meet each learning objective. After studying this module, you should be able to:

- explain the purpose of life-cycle analysis.
- describe alternative ways to handle waste and waste generation.

Preview Key Terms

In a notebook or on a separate sheet of paper, create a table like the one shown here to help with learning new key terms in the module. Before you read, fill out the "Prediction" column. Write what you think the term might mean or what it makes you think about. Use examples from your everyday life. There are no wrong answers!

Key Term	Prediction	Definition
Write key term here.	*Write what you think the term means in this column.*	*Define the term here. Add an example and use it in a sentence.*

Key Terms

Life-cycle analysis
Cradle-to-grave analysis
Integrated waste management

WHILE YOU READ THE MODULE

Define Key Terms

When you come across a new key term while reading the module, copy the definition into the "Definition" column of your key terms table. Add an example and use the term in a sentence. Compare your initial ideas to the actual definition.

AFTER YOU READ THE MODULE

Review Key Terms

Match the key terms on the left with the definitions on the right.

_____1. Life-cycle analysis (Cradle-to-grave analysis)

a. A systems tool that looks at the materials used and released throughout the lifetime of a product—from the procurement of raw materials through their manufacture, use, and disposal

_____2. Integrated waste management

b. An approach to waste disposal that employs several waste reduction, management, and disposal strategies in order to reduce the environmental impact of MSW

Chapter ⑯ Review Exercises

Check Your Understanding

Review "Learning Objectives Revisited" on page 584 of your textbook. Compare the notes you took while reading each module. Complete these exercises to review the chapter.

1. Summarize the composition of municipal solid waste in the United States.

2. List 5 parts of a sanitary landfill.

3. What are some problems with landfills?

4. What are some problems with incineration of waste?

5. Summarize the RCRA and CERCLA laws.

6. Explain Integrated Waste Management.

Practice for Free-Response Questions

Complete this exercise to build and practice the skills you will need to answer free-response questions on the exam. Use a separate sheet of paper if necessary.

What is e-waste? Why is it a growing environmental problem?

Review and Reflect

Complete these activities to solidify your knowledge of the chapter concepts and key terms. Use a notebook or a separate sheet of paper if necessary.

1. Review your key terms table for each module.

    (a) Which words did you already know? Which were new to you?
    (b) Write a new sentence using each key term.
    (c) Create a set of flash cards that includes each key term. Use the cards to review terms that were new or challenging.
    (d) When you feel comfortable with the new or challenging terms, review all of the cards, including those with familiar terms.
    (e) Save your cards to review before an exam.

2. What are the challenging concepts from this chapter?

    (a) Identify any concepts you found particularly challenging in this chapter.
    (b) Create a list of topics you need to review in preparation for an exam.

3. What questions do you have about concepts in the chapter?

    (a) Note any further questions you might have about material in the chapter.
    (b) Work with a partner to discuss these questions and ask your teacher for help as needed.

4. Write five possible multiple-choice questions based on this chapter. Work with a partner to quiz each other in preparation for an exam.

Chapter (17) Human Health and Environmental Risks

Chapter Summary

This chapter examines human health and environmental risk from the perspective of an environmental scientist. The chapter begins with a look at historical and emergent infectious diseases and then considers toxicology and chemical risks. It investigates chemicals of major concern, which are grouped into five categories: neurotoxins, carcinogens, teratogens, allergens, and endocrine disruptors. Other topics introduced include dose-response studies and biomagnification. The chapter concludes with an in-depth look at risk analysis and the three major steps in the risk-analysis process: risk assessment, risk acceptance, and risk management. Bioaccumulation and biomagnification are important topics to study, because they frequently appear on the exam. The chapter consists of 3 modules:

- **Module 56:** Human Diseases
- **Module 57:** Toxicology and Chemical Risks
- **Module 58:** Risk Analysis

Chapter Opening Case: *Citizen Scientists*

The chapter opening case introduces you to Margie Richard, winner of the 2004 Goldman Environmental Prize. Richard fought against pollution in her neighborhood, caused by a nearby oil refinery and chemical plants owned by Shell Oil Company. Richard didn't just organize community action – she learned how to gather scientific evidence to demonstrate the problems caused by pollution and to document the amount of toxic chemicals that were being released into the air. This case provides an example of how individual actions can make a difference.

Do the Math

This chapter contains the following "Do the Math" box to help prepare you for calculation questions you might encounter on the exam.

- "Estimating LD50 Values and Safe Exposures" (page 607)

To make sure you understand the concepts and techniques presented in this box, do the practice problems presented in the text as well as the additional "Practice the Math" problems that appear in Module 57 of this study guide.

Module 56: Environmental Science

BEFORE YOU READ THE MODULE

Focus on Learning Objectives

Use the module learning objectives to guide your reading. On a separate piece of paper, write down each objective and take notes to help you meet each learning objective. After studying this module, you should be able to:

- identify the different types of human diseases.
- understand the risk factors for human chronic diseases.
- discuss the historically important human diseases.
- identify the major emergent infectious diseases.
- discuss the future challenges for improving human health.

Preview Key Terms

In a notebook or on a separate sheet of paper, create a table like the one shown here to help with learning new key terms in the module. Before you read, fill out the "Prediction" column. Write what you think the term might mean or what it makes you think about. Use examples from your everyday life. There are no wrong answers!

Key Term	Prediction	Definition
Write key term here.	*Write what you think the term means in this column.*	*Define the term here. Add an example and use it in a sentence.*

Key Terms

Disease
Infectious disease
Acute disease
Chronic disease
Epidemic
Pandemic
Plague
Malaria
Tuberculosis
Emergent infectious disease

Acquired Immune Deficiency Syndrome (AIDS)
Human Immunodeficiency Virus (HIV)
Ebola hemorrhagic fever
Mad cow disease
Prion
Swine flu
Bird flu
Severe acute respiratory syndrome (SARS)
West Nile Virus

WHILE YOU READ THE MODULE

Define Key Terms

When you come across a new key term while reading the module, copy the definition into the "Definition" column of your key terms table. Add an example and use the term in a sentence. Compare your initial ideas to the actual definition.

Study the Figure

Use Figure 56.1, "Leading causes of death in the world" on page 592 to answer the following questions. Use a separate sheet of paper if necessary.

1. Which diseases account for over 50 percent of all deaths worldwide?

2. Among the world's death's caused by infectious diseases, 95 percent are caused by only 6 different types. What has helped eradicate many of these diseases?

Study the Figure

Use Figure 56.6, "Tuberculosis cases and deaths" on Page 595 to answer the following questions. Use a separate sheet of paper if necessary.

1. What is the approximate percent change of tuberculosis deaths worldwide from 2000 to 2010?

2. Why did tuberculosis deaths continue to decline in the United States even during the spike of cases worldwide in the years 1985-1995?

AFTER YOU READ THE MODULE

Review Key Terms

Match the key terms on the left with the definitions on the right.

_____1. Disease

_____2. Infectious disease

a. A type of flu caused by a coronavirus

b. An infectious disease caused by a bacterium *(Yersinia pestis)* that is carried by fleas

_____3. Acute disease

c. A disease that slowly impairs the functioning of an organism

_____4. Chronic disease

d. A disease that rapidly impairs the functioning of an organism

_____5. Epidemic

e. An infectious disease caused by the human immunodeficiency virus (HIV)

_____6. Pandemic

f. A situation in which a pathogen causes a rapid increase in disease

_____7. Plague

g. A disease in which prions mutate into deadly pathogens and slowly damage a cow's nervous system

_____8. Malaria

h. An infectious disease caused by one of several species of protists in the genus *Plasmodium*

_____9. Tuberculosis

i. A type of flu caused by the H1N1 virus

_____10. Emergent infectious disease

j. An infectious disease that has not been previously described or has not been common for at least 20 years

_____11. Acquired Immune Deficiency Syndrome (AIDS)

k. A disease caused by a pathogen

_____12. Human Immunodeficiency Virus (HIV)

l. A virus that lives in hundreds of species of birds and is transmitted among birds by mosquitoes

_____13. Ebola hemorrhagic fever

m. An epidemic that occurs over a large geographic region

_____14. Mad cow disease

n. A small, beneficial protein that occasionally mutates into a pathogen

_____15. Prion

o. An infectious disease with high death rates, caused by the Ebola virus

_____16. Swine flu

p. A type of flu caused by the H5N1 virus

_____17. Bird flu

q. Any impaired function of the body with a characteristic set of symptoms

_____18. Severe acute respiratory syndrome (SARS)

r. A type of virus that causes Acquired Immune Deficiency Syndrome (AIDS)

_____19. West Nile Virus

s. A highly contagious disease caused by the bacterium *Mycobacterium tuberculosis* that primarily infects the lungs

Module 57: Toxicology and Chemical Risks

BEFORE YOU READ THE MODULE

Focus on Learning Objectives

Use the module learning objectives to guide your reading. On a separate piece of paper, write down each objective and take notes to help you meet each learning objective. After studying this module, you should be able to:

- identify the major types of harmful chemicals.
- explain how scientists determine the concentrations of chemicals that harm organisms.

Preview Key Terms

In a notebook or on a separate sheet of paper, create a table like the one shown here to help with learning new key terms in the module. Before you read, fill out the "Prediction" column. Write what you think the term might mean or what it makes you think about. Use examples from your everyday life. There are no wrong answers!

Key Term	Prediction	Definition
Write key term here.	*Write what you think the term means in this column.*	*Define the term here. Add an example and use it in a sentence.*

Key Terms

Neurotoxin
Carcinogen
Mutagen
Teratogen
Allergen
Endocrine disruptor
Dose-response study

Acute study
Chronic study
LD50
Sublethal effect
ED50
Retrospective study
Prospective study

Synergistic interaction
Route of exposure
Solubility
Bioaccumulation
Biomagnification
Persistence

WHILE YOU READ THE MODULE

Define Key Terms

When you come across a new key term while reading the module, copy the definition into the "Definition" column of your key terms table. Add an example and use the term in a sentence. Compare your initial ideas to the actual definition.

Practice the Math: Estimating LD50 Values and Safe Exposures

Read "Do the Math: Estimating LD50 Values and Safe Exposures" on page 607. Try "Your Turn." For more math practice, do the following problem. Remember to show your work.

The LD50 of a particular pesticide for a rat is 2 mg/kg of mass. Assume that to determine the safe level of a pesticide for a dog you divide this LD50 value by 100. What amount of pesticide would be considered safe for a dog to ingest?

AFTER YOU READ THE MODULE

Review Key Terms

Match the key terms on the left with the definitions on the right.

_____1. Neurotoxin

_____2. Carcinogen

_____3. Mutagen

_____4. Teratogen

_____5. Allergen

_____6. Endocrine disruptor

_____7. Dose-response study

_____8. Acute study

a. The lethal dose of a chemical that kills 50 percent of the individuals in a dose-response study

b. A situation in which two risks together cause more harm than expected based on the separate effects of each risk alone

c. A chemical that causes allergic reactions.

d. An increased concentration of a chemical within an organism over time

e. The length of time a chemical remains in the environment

f. How well a chemical dissolves in a liquid

g. A chemical that interferes with the normal functioning of hormones in an animal's body

h. The effective dose of a chemical that causes 50 percent of the individuals in a dose-response study to display a harmful, but nonlethal, effect

_____9. Chronic study

_____10. LD50

_____11. Sublethal effect

_____12. ED50

_____13. Retrospective study

_____14. Prospective study

_____15. Synergistic interaction

_____16. Route of exposure

_____17. Solubility

_____18. Bioaccumulation

_____19. Biomagnification

_____20. Persistence

i. A study that monitors people who have been exposed to an environmental hazard at some time in the past

j. A chemical that interferes with the normal development of embryos or fetuses

k. A study that monitors people who might become exposed to harmful chemicals in the future

l. The increase in chemical concentration in animal tissues as the chemical moves up the food chain

m. A study that exposes organisms to different amounts of a chemical and then observes a variety of possible responses, including mortality or changes in behavior or reproduction

n. A chemical that disrupts the nervous systems of animals

o. The effect of an environmental hazard that is not lethal, but which may impair an organism's behavior, physiology, or reproduction

p. An experiment that exposes organisms to an environmental hazard for a long duration

q. The way in which an individual might come into contact with an environmental hazard

r. A chemical that causes cancer

s. An experiment that exposes organisms to an environmental hazard for a short duration

t. A type of carcinogen that causes damage to the genetic material of a cell

Module 58: Scientific Method

Focus on Learning Objectives

Use the module learning objectives to guide your reading. On a separate piece of paper, write down each objective and take notes to help you meet each learning objective. After studying this module, you should be able to:

- explain the processes of qualitative versus quantitative risk assessment.
- understand how to determine the amount of risk that can be tolerated.
- discuss how risk management balances potential harm against other factors.
- contrast the innocent-until-proven-guilty principle and the precautionary principle.

Preview Key Terms

In a notebook or on a separate sheet of paper, create a table like the one shown here to help with learning new key terms in the module. Before you read, fill out the "Prediction" column. Write what you think the term might mean or what it makes you think about. Use examples from your everyday life. There are no wrong answers!

Key Term	Prediction	Definition
Write key term here.	*Write what you think the term means in this column.*	*Define the term here. Add an example and use it in a sentence.*

Key Terms

Environmental hazard
Innocent-until-proven-guilty principle
Precautionary principle

Stockholm Convention
REACH

WHILE YOU READ THE MODULE

Define Key Terms

When you come across a new key term while reading the module, copy the definition into the "Definition" column of your key terms table. Add an example and use the term in a sentence. Compare your initial ideas to the actual definition.

Review Key Terms

Match the key terms on the left with the definitions on the right.

_____1. Environmental hazard

a. A principle based on the belief that a potential hazard should not be considered an actual hazard until the scientific data definitively demonstrate that it actually causes harm

_____2. Innocent-until-proven-guilty principle

b. A 2007 agreement among the nations of the European Union about regulation of chemicals; the acronym stands for registration, evaluation, authorization, and restriction of chemicals

_____3. Precautionary principle

c. Anything in the environment that can potentially cause harm

_____4. Stockholm Convention

d. A 2001 agreement among 127 nations concerning 12 chemicals to be banned, phased out, or reduced

_____5. REACH

e. A principle based on the belief that action should be taken against a plausible environmental hazard

Chapter (17) Review Exercises

Check Your Understanding

Review "Learning Objectives Revisited" on page 620 of your textbook. Compare the notes you took while reading each module. Complete these exercises to review the chapter.

1. What are some health risks for people living in a developing nation? A developed nation?

2. Fill out the following chart:

Disease	How Disease is Spread
Plague	
Malaria	
Tuberculosis	
HIV/AIDS	
Ebola hemorrhagic fever	
Mad cow disease	
Bird flu	
West Nile virus	

3. What are the different pathways of transmitting pathogens?

4. Fill out the following chart:

Chemical	Source	Type	Effect
Lead			
Mercury			
Arsenic			
Asbestos			
PCBs			
Radon			
Vinyl chloride			
Alcohol			
Atrazine			
DDT			
Phthalates			

5. Explain the difference between LD50 and ED50.

6. How did DDT move up the food chain and harm predatory birds?

7. What factors are included in risk analysis?

Practice for Free-Response Questions

Complete this exercise to build and practice the skills you will need to answer free-response questions on the exam. Use a separate sheet of paper if needed.

Explain the difference between bioaccumulation and biomagnification and give examples of each.

Review and Reflect

Complete these activities to solidify your knowledge of the chapter concepts and key terms. Use a notebook or a separate sheet of paper if necessary.

1. Review your key terms table for each module.

   (a) Which words did you already know? Which were new to you?
   (b) Write a new sentence using each key term.
   (c) Create a set of flash cards that includes each key term. Use the cards to review terms that were new or challenging.
   (d) When you feel comfortable with the new or challenging terms, review all of the cards, including those with familiar terms.
   (e) Save your cards to review before an exam.

2. What are the challenging concepts from this chapter?

   (a) Identify any concepts you found particularly challenging in this chapter.
   (b) Create a list of topics you need to review in preparation for an exam.

3. What questions do you have about concepts in the chapter?

   (a) Note any further questions you might have about material in the chapter.
   (b) Work with a partner to discuss these questions and ask your teacher for help as needed.

4. Write five possible multiple-choice questions based on this chapter. Work with a partner to quiz each other in preparation for an exam.

Unit 7 Multiple-Choice Review Exam

Choose the best answer.

1. Which is an example of a point source water pollutant?
   (A) Agricultural lands
   (B) Animal feedlots
   (C) Runoff from parking lots
   (D) Factory effluent
   (E) Residential lawns

2. When sewage contaminates a body of water, it can lead to a lower dissolved oxygen level in the water because
   (A) the sewage has put a high BOD on the water.
   (B) the water has too many fish and other organisms.
   (C) the sewage magnifies existing toxins.
   (D) of eutrophication.
   (E) the lake is oligotrophic.

3. The World Health Organization estimates that _____ people does not have access to sufficient supplies of safe drinking water.
   (A) 1 out of every 100
   (B) 1 out of every 50
   (C) 1 out of every 25
   (D) 1 out of every 10
   (E) 1 out of every 6

Match the following steps in the sewage treatment process.

4. _____ Biological   (A) Solid waste materials settle out
5. _____ Chemical   (B) Bacteria break down 85-90 percent of organic matter
6. _____ Mechanical   (C) Chlorine, ozone, or ultraviolet light are used

7. Which is NOT a primary pollutant?
   (A) Carbon monoxide
   (B) Ozone
   (C) Carbon dioxide
   (D) Sulfur dioxide
   (E) Volatile organic compounds

8. Which is a corrosive gas that comes primarily from burning coal?
(A) Carbon monoxide
(B) Ozone
(C) Carbon dioxide
(D) Sulfur dioxide
(E) Methane

9. Which causes acid deposition?
   I. Sulfur dioxide
   II. Nitrogen oxides
   III. Carbon dioxide
(A) I only
(B) II only
(C) III only
(D) I and II
(E) I, II, and III

10. In a coal burning power plant, an electrostatic precipitator is designed to remove
(A) sulfur dioxide.
(B) nitrogen dioxide
(C) particulate matter.
(D) carbon dioxide.
(E) methane.

11. A thermal inversion causes severe pollution events. Thermal inversions
(A) occur when a warm inversion layer traps emissions that then accumulate beneath it.
(B) occur when a cold inversion layer traps emissions that then accumulate beneath it.
(C) occur when a warm inversion layer traps emissions that then accumulate above it.
(D) occur when cold inversion layer traps emissions that then accumulate above it.
(E) occur mostly in mid-latitudes.

12. Which of the following statements about stratospheric ozone depletion is correct?
(A) Ozone depletion is the result of automobiles and coal fired power plants.
(B) Ozone depletion occurs when infrared heat is trapped near the earth.
(C) Ozone depletion is causing melting of our polar ice caps.
(D) Ozone depletion is causing respiratory disease to increase.
(E) Ozone depletion allows more ultraviolet waves to pass through to the troposphere.

13. Which chemical damages the stratospheric ozone layer?
(A) Carbon
(B) Chlorine
(C) Methane
(D) Sulfur dioxide
(E) Fluorine

14. Which makes up most of the municipal solid waste stream in the United States?
   (A) Paper
   (B) Plastic
   (C) Yard trimmings
   (D) Food scraps
   (E) Wood

15. Which uses the least amount of energy?
   (A) Reduce
   (B) Reuse
   (C) Closed-loop recycling
   (D) Open-loop recycling
   (E) Incineration

Use the following graph to answer question 16.

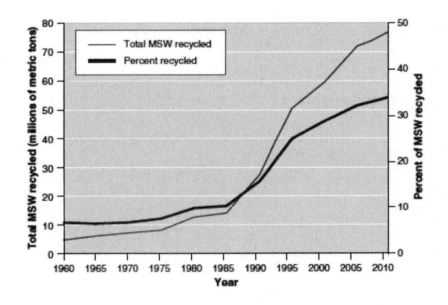

16. Using the graph above, what is the approximate percent change of materials that were recycled in 2008 compared to 1960?
   (A) 10 percent
   (B) 25 percent
   (C) 75 percent
   (D) 100 percent
   (E) 380 percent

17. Which is NOT a consequence of incineration?
   (A) Production of fly ash
   (B) Fossil fuels burned to transport waste
   (C) Production of leachate
   (D) Production of air pollution
   (E) Release of metals and other toxins

18. Which law was created in part because of Love Canal?
    (A) RCRA
    (B) Brownfields
    (C) ESA
    (D) CERCLA
    (E) FIFRA

19. Integrated waste management employs
        I. source reduction.
        II. recycling.
        III. composting.
    (A) I only
    (B) II only
    (C) III only
    (D) I and II only
    (E) I, II, and III

20. The leading health risk in high-income countries is
    (A) automobiles.
    (B) tobacco use.
    (C) obesity.
    (D) low fruit and vegetable intake.
    (E) high cholesterol.

21. Which is NOT an infectious disease?
    (A) Cancer
    (B) Plague
    (C) Tuberculosis
    (D) HIV/AIDS
    (E) Ebola

22. Which is a neurotoxin?
    (A) Arsenic
    (B) Asbestos
    (C) Lead
    (D) Synthetic hormones
    (E) Radon

23. Which experiment results correctly describes the effects of a LD50 dose of a chemical?
    (A) One out of every 50 rats die.
    (B) Fifty out of every 100 rats die.
    (C) One out of every 50 rats gets sick.
    (D) Fifty out of every 100 rats get sick.
    (E) All rats die.

24. Which would be an example of a synergistic interaction?
    (A) Increased riders in a car leads to increased deaths.
    (B) Smoking causes lung disease.
    (C) Tuberculosis is spread among the poor.
    (D) Recreational drug use can lead to HIV/AIDS.
    (E) A smoker who is also exposed to asbestos is more likely to get lung cancer.

25. The chemical DDT consumed by plankton is transferred to small fish, to larger fish, and
    eventually to predatory birds. This is an example of
    (A) a food chain.
    (B) a synergistic effect.
    (C) risk assessment.
    (D) biomagnification.
    (E) teratogens.

26. A scientist is studying a local river and notices that many of the male fish, reptiles, and
    amphibians are becoming feminized; males possess testes that have low sperm counts, and in
    some cases, testes that produce both eggs and sperm. Based on this evidence, what might the
    scientist suspect?
    (A) The water contains paint from a nearby industrial site.
    (B) The pH of the river has changed because of acid precipitation.
    (C) A local factory has been dumping wastewater with high levels of PCBs.
    (D) The river contains a large quantity of gasoline runoff from a local gas station.
    (E) The organisms have been exposed to wastewater containing pharmaceutical drugs that
    mimic estrogen.

27. Which location would most likely experience photochemical smog?
    (A) Oslo, Norway
    (B) Los Angeles, California
    (C) Manaus, Brazil
    (D) Vancouver, British Columbia
    (E) London, England

28. A young child has been exposed to a particular chemical compound and shows signs of
    central nervous system damage. Which of the following chemicals might you suspect?
    (A) Ozone
    (B) Carbon monoxide
    (C) Sulfur dioxide
    (D) Volatile organic compounds
    (E) Lead

Chapter (18) Conservation of Biodiversity

Chapter Summary

This chapter explores the reasons for the decline in biodiversity and various strategies for preserving biodiversity. The chapter begins by noting that extinctions are a natural process and have occurred regularly throughout Earth's history. However, in recent times the number of extinctions has increased as the result of human activity. The chapter goes on to look at the specific mechanisms for the loss of biodiversity, and finally examines efforts currently underway to conserve biodiversity. The chapter consists of 3 modules:

- **Module 59:** The Sixth Mass Extinction
- **Module 60:** Causes of Declining Biodiversity
- **Module 61:** The Conservation of Biodiversity

Chapter Opening Case: *Modern Conservation Legacies*

The chapter opening case introduces you to some of the important conservation work being done today. It describes the Papaha–naumokua–kea Marine National Monument and other areas set aside for the protection and enjoyment of the natural world. This case provides a good opportunity for you to discuss the different levels of conservation, from genetic diversity to habitat preservation. It is also important for you to understand the economic pressures involved in setting aside land and resources for protection as well as the economic benefits of preserving biodiversity. Past free-response questions have asked students to write about both the ecological and the economic pros and cons of species preservation.

Module 59: The Sixth Mass Extinction

BEFORE YOU READ THE MODULE

Focus on Learning Objectives

Use the module learning objectives to guide your reading. On a separate piece of paper, write down each objective and take notes to help you meet each learning objective. After studying this module, you should be able to:

- explain the global decline in the genetic diversity of wild species.
- discuss the global decline in the genetic diversity of domesticated species.
- identify the patterns of global decline in species diversity.
- explain the values of ecosystems and the global declines in ecosystem function.

Preview Key Terms

In a notebook or on a separate sheet of paper, create a table like the one shown here to help with learning new key terms in the module. Before you read, fill out the "Prediction" column. Write what you think the term might mean or what it makes you think about. Use examples from your everyday life. There are no wrong answers!

Key Term	Prediction	Definition
Write key term here.	*Write what you think the term means in this column.*	*Define the term here. Add an example and use it in a sentence.*

Key Terms

Threatened species
Near-threatened species
Least concern species

Intrinsic value
Instrumental value
Instrumental value

Provision

WHILE YOU READ THE MODULE

Define Key Terms

When you come across a new key term while reading the module, copy the definition into the "Definition" column of your key terms table. Add an example and use the term in a sentence. Compare your initial ideas to the actual definition.

Study the Figure

Examine Figure 59.4, "The decline of birds, mammals, and amphibians" on page 636.

1. Which group of species has had the least percentage of global declines since the year 1500?

AFTER YOU READ THE MODULE

Review Key Terms

Match the key terms on the left with the definitions on the right.

_____1. Threatened species

a. Species that are very likely to become threatened in the future

_____2. Near-threatened species

b. Worth as an instrument or a tool that can be used to accomplish a goal

_____3. Least concern species

c. Worth as an instrument or a tool that can be used to accomplish a goal

_____4. Intrinsic value

   d. According to the International Union for Conservation of Nature (IUCN), species that have a high risk of extinction in the future

_____5. Instrumental value

   e. A good that humans can use directly

_____6. Instrumental value

   f. Species that are widespread and abundant

_____7. Provision

   g. Value independent of any benefit to humans

Module 60: Causes of Declining Biodiversity

BEFORE YOU READ THE MODULE

Focus on Learning Objectives

Use the module learning objectives to guide your reading. On a separate piece of paper, write down each objective and take notes to help you meet each learning objective. After studying this module, you should be able to:

- discuss how habitat loss can lead to declines in species diversity.
- explain how the movement of exotic species affects biodiversity.
- describe how overharvesting causes declines in populations and species.
- understand how pollution reduces populations and biodiversity.
- identify how climate change affects species diversity.

Preview Key Terms

In a notebook or on a separate sheet of paper, create a table like the one shown here to help with learning new key terms in the module. Before you read, fill out the "Prediction" column. Write what you think the term might mean or what it makes you think about. Use examples from your everyday life. There are no wrong answers!

Key Term	Prediction	Definition
Write key term here.	_Write what you think the term means in this column._	_Define the term here. Add an example and use it in a sentence._

Key Terms

Ecosystem service
Environmental indicator
Biodiversity
Genetic
Diversity
Species

Species
diversity
Speciation
Background
Extinction rate
Greenhouse gases

Anthropogenic
Development
Sustainability
Sustainable development
Biophilia
Ecological footprint

Define Key Terms

When you come across a new key term while reading the module, copy the definition into the "Definition" column of your key terms table. Add an example and use the term in a sentence. Compare your initial ideas to the actual definition.

Study the Figure

Examine Figure 60.2, "Changing forests" on page 643.

1. What portion of the world has seen the greatest increase in forest cover?

AFTER YOU READ THE MODULE

Review Key Terms

Match the Key Terms on the left with the definitions on the right.

_____1. Native species

a. A species that spreads rapidly across large areas

_____2. Exotic species (alien species)

b. A species living outside its historical range

_____3. Invasive species

c. A 1973 treaty formed to control the international trade of threatened plants and animals

_____4. Lacey Act

d. A list of worldwide threatened species

_____5. Convention on International Trade in Endangered Species of Wild Fauna and Flora (CITES)

e. Species that live in their historical range, typically where they have lived for thousands or millions of years

_____6. Red List

f. A U.S. act that prohibits interstate shipping of all illegally harvested plants and animals

Module 61: The Conservation of Biodiversity

BEFORE YOU READ THE MODULE

Focus on Learning Objectives

Use the module learning objectives to guide your reading. On a separate piece of paper, write down each objective and take notes to help you meet each learning objective. After studying this module, you should be able to:

- identify legislation that focuses on protecting single species.
- discuss conservation efforts that focus on protecting entire ecosystems.

Preview Key Terms

In a notebook or on a separate sheet of paper, create a table like the one shown here to help with learning new key terms in the module. Before you read, fill out the "Prediction" column. Write what you think the term might mean or what it makes you think about. Use examples from your everyday life. There are no wrong answers!

Key Term	Prediction	Definition
Write key term here.	Write what you think the term means in this column.	Define the term here. Add an example and use it in a sentence.

Key Terms

Marine Mammal Act
Endangered species
Threatened species

Convention on Biological Diversity
Edge habitat
Biosphere reserve

WHILE YOU READ THE MODULE

Define Key Terms

When you come across a new key term while reading the module, copy the definition into the "Definition" column of your key terms table. Add an example and use the term in a sentence. Compare your initial ideas to the actual definition.

Review Key Terms

Match the key terms on the left with the definitions on the right.

_____1. Marine Mammal Act

                 a. A 1972 U.S. act to protect declining populations of marine mammals

_____2. Endangered species

                 b. According to U.S. legislation, any species that is likely to become an endangered species within the foreseeable future throughout all or a significant portion of its range

_____3. Threatened species

                 c. A species that is in danger of extinction within the foreseeable future throughout all or a significant portion of its range

_____4. Convention on Biological Diversity

                 d. Protected area consisting of zones that vary in the amount of permissible human impact

_____5. Edge habitat

                 e. An international treaty to help protect biodiversity

_____6. Biosphere reserve

                 f. Habitat that occurs where two different communities come together, typically forming an abrupt transition, such as where a grassy field meets a forest

Chapter (18) Review Exercises

Check Your Understanding

Review "Learning Objectives Revisited" on page 659 of your textbook. Compare the notes you took while reading each module. Complete these exercises to review the chapter.

    1. What are the five categories used by the International Union for Conservation of Nature to identify the status of a species?

2. How do invasive species threaten biodiversity? Give an example of an invasive species.

3. Summarize the aims of following laws and treaties.

   Lacey Act
   CITES
   Marine Mammal Protection Act
   Endangered Species Act
   Convention on Biological Diversity

4. Describe a biosphere reserve.

5. What are some of the reasons behind biodiversity decline?

6. How can we reduce species decline?

Practice for Free-Response Questions

Complete this exercise to build and practice the skills you will need to answer free-response questions on the exam. Use a separate sheet of paper if necessary.

What is the 2002 Convention on Biological Diversity? Identify two trends that the 2002 Convention on Biological Diversity recognized.

Review and Reflect

Complete these activities to solidify your knowledge of the chapter concepts and key terms. Use a notebook or a separate sheet of paper if necessary.

1. Review your key terms table for each module.

    (a) Which words did you already know? Which were new to you?
    (b) Write a new sentence using each key term.
    (c) Create a set of flash cards that includes each key term. Use the cards to review terms that were new or challenging.
    (d) When you feel comfortable with the new or challenging terms, review all of the cards, including those with familiar terms.
    (e) Save your cards to review before an exam.

2. What are the challenging concepts from this chapter?

    (a) Identify any concepts you found particularly challenging in this chapter.
    (b) Create a list of topics you need to review in preparation for an exam.

3. What questions do you have about concepts in the chapter?

    (a) Note any further questions you might have about material in the chapter.
    (b) Work with a partner to discuss these questions and ask your teacher for help as needed.

4. Write five possible multiple-choice questions based on this chapter. Work with a partner to quiz each other in preparation for an exam.

Chapter (19) Global Change

Chapter Summary

This chapter examines how humans have altered the world's climate. It explores the underlying causes of global change and their consequences. In this chapter, students have the opportunity to apply many of the themes developed throughout the course, including the interconnectedness of the systems on Earth, environmental indicators, and the interaction of environmental science and policy. This chapter provides several opportunities for reviewing the biogeochemical cycles covered in Chapter 3. The chapter consists of 3 modules:

- **Module 62:** Global Climate Change and the Greenhouse Effect
- **Module 63:** The Evidence for Global Warming
- **Module 64:** The Consequences of Global Climate Change

Chapter Opening Case: *Walking on Thin Ice*

The chapter opening case introduces students to the themes of this chapter using the example of polar bears. Earth's climate is a global system and the actions of one nation can affect climate conditions many thousands of miles away. Rising temperatures in the Arctic have caused the polar ice cap to melt, leading to a loss of the polar bear's habitat. A decline in polar bear population could have extremely negative consequences for the entire ecosystem.

Do the Math

This chapter contains the following "Do the Math" box to help prepare you for calculation questions you might encounter on the exam.

- "Projecting Future Increases in CO_2" (page 675)

To make sure you understand the concepts and techniques presented in this box, do the practice problems presented in the text as well as the additional "Practice the Math" problems that appear in Module 63 of this study guide.

Module 62: Global Climate Change and the Greenhouse Effect

BEFORE YOU READ THE MODULE

Focus on Learning Objectives

Use the module learning objectives to guide your reading. On a separate piece of paper, write down each objective and take notes to help you meet each learning objective. After studying this module, you should be able to:

- distinguish among global change, global climate change, and global warming.
- explain the process underlying the greenhouse effect.
- identify the natural and anthropogenic sources of greenhouse gases.

Preview Key Terms

In a notebook or on a separate sheet of paper, create a table like the one shown here to help with learning new key terms in the module. Before you read, fill out the "Prediction" column. Write what you think the term might mean or what it makes you think about. Use examples from your everyday life. There are no wrong answers!

Key Term	Prediction	Definition
Write key term here.	*Write what you think the term means in this column.*	*Define the term here. Add an example and use it in a sentence.*

Key Terms

Global change	Global warming	Greenhouse warming potential
Global climate change	Greenhouse effect	

WHILE YOU READ THE MODULE

Define Key Terms

When you come across a new key term while reading the module, copy the definition into the "Definition" column of your key terms table. Add an example and use the term in a sentence. Compare your initial ideas to the actual definition.

Study The Figure

Use Figure 62.2, "The greenhouse effect" on page 667 to answer the following questions.

1. Label the steps of the greenhouse effect.

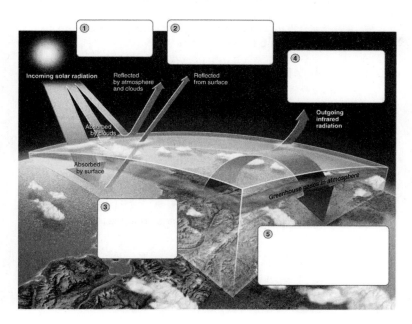

2. Describe two environmental implications that occur when chemicals build up in our atmosphere and trap more heat from the sun.

Study The Figure

Use Figure 62.6, "Anthropogenic sources of greenhouse gases in the United States" on page 672 to answer the following questions.

1. What is responsible for the largest contribution of methane?

2. 69 percent of nitrogenous oxide pollution comes from agriculture soils. What anthropogenic cause leads to nitrous oxide in the soils?

3. What two anthropogenic sources account for over 60 percent of carbon dioxide in our atmosphere?

Review Key Terms

Match the key terms on the left with the definitions on the right.

_____1. Global change

a. Changes in the average weather that occurs in an area over a period of years or decades

_____2. Global climate change

b. Absorption of infrared radiation by atmospheric gases and reradiation of the energy back toward Earth

_____3. Global warming

c. Change that occurs in the chemical, biological, and physical properties of the planet

_____4. Greenhouse effect

d. An estimate of how much a molecule of any compound can contribute to global warming over a period of 100 years relative to a molecule of CO_2

_____5. Greenhouse warming potential

e. The warming of the oceans, land masses, and atmosphere of Earth

Module 63: The Evidence

BEFORE YOU READ THE MODULE

Focus on Learning Objectives

Use the module learning objectives to guide your reading. On a separate piece of paper, write down each objective and take notes to help you meet each learning objective. After studying this module, you should be able to:

- identify key environmental indicators and their trends over time.
- define sustainability and explain how it can be measured using the ecological footprint.

Preview Key Terms

In a notebook or on a separate sheet of paper, create a table like the one shown here to help with learning new key terms in the module. Before you read, fill out the "Prediction" column. Write what you think the term might mean or what it makes you think about. Use examples from your everyday life. There are no wrong answers!

Key Term	Prediction	Definition
Write key term here.	*Write what you think the term means in this column.*	*Define the term here. Add an example and use it in a sentence.*

Key Terms

Ocean acidification

WHILE YOU READ THE MODULE

Define Key Terms

When you come across a new key term while reading the module, copy the definition into the "Definition" column of your key terms table. Add an example and use the term in a sentence. Compare your initial ideas to the actual definition.

Study the Figure

Use Figure 63.1, " Changes in atmospheric CO_2 over time" on page 675 to answer the following questions.

1. What was the percent change of atmospheric CO_2 between 1994 and 2004?

2. Identify two reasons for the steady rise in atmospheric CO_2.

Practice the Math: Projecting Future Increases in CO₂

Read "Do the Math: Projecting Future Increases in CO_2" on page 675. Try "Your Turn." For more math practice, do the problems below. Remember to show your work.

1. In 2010, the concentration of carbon dioxide in the atmosphere was 390 ppm (parts per million). If the annual rate of carbon dioxide increase is 1.4 ppm, what concentration of carbon dioxide do you predict for the year 2050?

2. If the annual rate of carbon dioxide increases from 1.4 ppm to 1.9 ppm, what will the concentration of carbon dioxide be in the year 2050?

3. From 1960 to 2010, the concentration of CO_2 in the atmosphere increased from 320 to 390 ppm. Using 2010 as your starting point, if the annual rate of CO_2 increase is 1.4 ppm, what concentration of CO_2 do you predict for the year 2150?

Module 64: Consequences of Global Climate Change

BEFORE YOU READ THE MODULE

Focus on Learning Objectives

Use the module learning objectives to guide your reading. On a separate piece of paper, write down each objective and take notes to help you meet each learning objective. After studying this module, you should be able to:

- discuss how global climate change has affected the environment.
- explain how global climate change has affected organisms.
- identify the future changes predicted to occur with global climate change.
- explain the global climate change goals of the Kyoto Protocol.

Preview Key Terms

In a notebook or on a separate sheet of paper, create a table like the one shown here to help with learning new key terms in the module. Before you read, fill out the "Prediction" column. Write what you think the term might mean or what it makes you think about. Use examples from your everyday life. There are no wrong answers!

Key Term	Prediction	Definition
Write key term here.	*Write what you think the term means in this column.*	*Define the term here. Add an example and use it in a sentence.*

Key Terms

Kyoto protocol
Carbon sequestration

WHILE YOU READ THE MODULE

Define Key Terms

When you come across a new key term while reading the module, copy the definition into the "Definition" column of your key terms table. Add an example and use the term in a sentence. Compare your initial ideas to the actual definition.

AFTER YOU READ THE MODULE

Review Key Terms

Match the key terms on the left with the definitions on the right.

_____1. Kyoto protocol

_____2. Carbon sequestration

a. An approach to stabilizing greenhouse gases by removing CO_2 from the atmosphere

b. An international agreement that sets a goal for global emissions of greenhouse gases from all industrialized countries to be reduced by 5.2 percent below their 1990 levels by 2012

Chapter ⑲ Review Exercises

Check Your Understanding

Review "Learning Objectives Revisited" on page 697 of your textbook. Compare the notes you took while reading each module. Complete these exercises to review the chapter.

1. Summarize the greenhouse effect.

2. List the five greenhouse gases and their relative impact on the greenhouse effect. Include their global warming potential and how long they remain in the atmosphere.

3. Why do carbon dioxide levels change seasonally?

4. How have scientists used ice cores to determine past climate patterns? How can these patterns help scientists make predictions about future climate change?

5. What are two explanations for warming temperatures on Earth?

6. Explain carbon sequestration.

Practice for Free-Response Questions

Complete this exercise to build and practice the skills you will need to answer free-response questions on the exam. Use a separate sheet of paper if necessary.

Describe how volcanic eruptions can affect the climate.

Review and Reflect

Complete these activities to solidify your knowledge of the chapter concepts and key terms. Use a notebook or a separate sheet of paper if necessary.

1. Review your key terms table for each module.

   (a) Which words did you already know? Which were new to you?
   (b) Write a new sentence using each key term.
   (c) Create a set of flash cards that includes each key term. Use the cards to review terms that were new or challenging.
   (d) When you feel comfortable with the new or challenging terms, review all of the cards, including those with familiar terms.
   (e) Save your cards to review before an exam.

2. What are the challenging concepts from this chapter?

   (a) Identify any concepts you found particularly challenging in this chapter.
   (b) Create a list of topics you need to review in preparation for an exam.

3. What questions do you have about concepts in the chapter?

   (a) Note any further questions you might have about material in the chapter.
   (b) Work with a partner to discuss these questions and ask your teacher for help as needed.

4. Write five possible multiple-choice questions based on this chapter. Work with a partner to quiz each other in preparation for an exam.

Chapter 20 Sustainability, Economics, and Equity

Chapter Summary

This chapter explores sustainability in the context of economics. It describes how sustainability can be achieved through the use of sound economic and business practices as well as through effective environmental regulations and laws. The chapter looks at the work of agencies such as the United Nations Environment Programme as well as a number of U.S. agencies, including the Environmental Protection Agency that oversees all governmental efforts related to the environment. The chapter describes the role of different worldviews in approaches to sustainability. Finally, the chapter considers the importance of reducing poverty as part of achieving sustainability. The chapter consists of 2 modules:

- **Module 65:** Sustainability and Economics
- **Module 66:** Regulations and Equity

Chapter Opening Case: *Assembly Plants, Free Trade, and Sustainable Systems*

The chapter opening case introduces you to the economic, social, and environmental implications of NAFTA, the international trade agreement with the United States, Mexico, and Canada. Students will consider the benefits to all three countries, the environmental and social consequences for Mexican community members, and the balance that must exist among economic profit, environmental integrity, and human welfare in order to establish sustainable development.

Module 65: Sustainability and Economics

BEFORE YOU READ THE MODULE

Focus on Learning Objectives

Use the module learning objectives to guide your reading. On a separate piece of paper, write down each objective and take notes to help you meet each learning objective. After studying this module, you should be able to:

- environmental science and economic analysis.
- describe how economic health depends on the availability of natural capital and basic human welfare.

Preview Key Terms

In a notebook or on a separate sheet of paper, create a table like the one shown here to help with learning new key terms in the module. Before you read, fill out the "Prediction" column. Write what you think the term might mean or what it makes you think about. Use examples from your everyday life. There are no wrong answers!

Key Term	Prediction	Definition
Write key term here.	*Write what you think the term means in this column.*	*Define the term here. Add an example and use it in a sentence.*

Key Terms

Well-being
Economics
Genuine progress indicator
(GPI)
Technology transfer

Leapfrogging
Natural capital
Human capital
Manufactured capital
Market failure

Environmental economics
Ecological economics
Valuation

WHILE YOU READ THE MODULE

Define Key Terms

When you come across a new key term while reading the module, copy the definition into the "Definition" column of your key terms table. Add an example and use the term in a sentence. Compare your initial ideas to the actual definition.

Study the Figure

Examine Figure 65.1, "Supply and demand" on page 704.

    1. What does the equilibrium point represent?

Study the Figure

Using Figure 65.4, "The Kuznets curve" on page 706, answer the following questions.

    1. As GDP initially increases, what happens to GPI?

    2. As a country develops, what are some things that could be put in place to improve the environment?

AFTER YOU READ THE MODULE

Review Key Terms

Match the key terms on the left with the definitions on the right.

_____1. Well-being

_____2. Economics

_____3. Genuine progress indicator (GPI)

_____4. Technology transfer

_____5. Leapfrogging

_____6. Natural capital

_____7. Human capital

_____8. Manufactured capital

_____9. Market failure

_____10. Environmental economics

_____11. Ecological economics

_____12. Valuation

a. The phenomenon of less developed countries adopting technological innovations developed in wealthy countries

b. When the economic system does not account for all costs

c. The study of economics as a component of ecological systems

d. The phenomenon of less developed countries using new technology without first using the precursor technology

e. Human knowledge and abilities

f. The study of how humans allocate scarce resources in the production, distribution, and consumption of goods and services

g. The practice of assigning monetary value to intangible benefits and natural capital

h. A measure of economic status that includes personal consumption, income distribution, levels of higher education, resource depletion, pollution, and the health of the population

i. The status of being healthy, happy, and prosperous

j. All goods and services that humans produce

k. A subfield of economics that examines the costs and benefits of various policies and regulations that seek to regulate or limit air and water pollution and other causes of environmental degradation

l. The resources of the planet, such as air, water, and minerals

Module 66: Regulations and Equity

BEFORE YOU READ THE MODULE

Focus on Learning Objectives

Use the module learning objectives to guide your reading. On a separate piece of paper, write down each objective and take notes to help you meet each learning objective. After studying this module, you should be able to:

- explain the role of agencies and regulations in efforts to protect our natural and human capital.
- describe the approaches to measuring and achieving sustainability.
- discuss the relationship among sustainability, poverty, personal action, and stewardship.

Preview Key Terms

In a notebook or on a separate sheet of paper, create a table like the one shown here to help with learning new key terms in the module. Before you read, fill out the "Prediction" column. Write what you think the term might mean or what it makes you think about. Use examples from your everyday life. There are no wrong answers!

Key Term	Prediction	Definition
Write key term here.	*Write what you think the term means in this column.*	*Define the term here. Add an example and use it in a sentence.*

Key Terms

Environmental worldview
Anthropocentric worldview
Stewardship
Biocentric worldview
Ecocentric worldview
United Nations (UN)
United Nations Environment Programme (UNEP)
World Bank
World Health Organization (WHO)
United Nations Development Programme (UNDP)

Environmental Protection Agency (EPA)
Occupational Safety and Health Administration (OSHA)
Department of Energy (DOE)
Human development index (HDI)
Human poverty index (HPI)
Command-and-control approach
Incentive-based approach
Green tax
Triple bottom line

WHILE YOU READ THE MODULE

Define Key Terms

When you come across a new key term while reading the module, copy the definition into the "Definition" column of your key terms table. Add an example and use the term in a sentence. Compare your initial ideas to the actual definition.

AFTER YOU READ THE MODULE

Review Key Terms

Match the key terms on the left with the definitions on the right.

_____1. Environmental worldview

    a. A worldview that places equal value on all living organisms and the ecosystems in which they live

_____2. Anthropocentric worldview

    b. A global institution dedicated to promoting dialogue among countries with the goal of maintaining world peace

_____3. Stewardship

    c. A global institution dedicated to the improvement of human health by monitoring and assessing health trends and providing medical advice to countries

_____4. Biocentric worldview

    d. A worldview that encompasses how one thinks the world works; how one views one's role in the world; and what one believes to be proper environmental behavior

_____5. Ecocentric worldview

    e. A program of the United Nations responsible for gathering environmental information, conducting research, and assessing environmental problems

_____6. United Nations (UN)

    f. The U.S. organization that advances the energy and economic security of the United States

_____7. United Nations Environment Programme (UNEP)

    g. A worldview that focuses on human welfare and well-being

_____8. World Bank

    h. An approach to sustainability that considers three factors—economic, environmental, and social—when making decisions about business, the economy, and development

_____9. World Health Organization (WHO)

    i. An international program that works in 166 countries around the world to advocate change that will help people obtain a better life through development

_____10. United Nations Development Programme (UNDP)

_____11. Environmental Protection Agency (EPA)

_____12. Occupational Safety and Health Administration (OSHA)

_____13. Department of Energy (DOE)

_____14. Human development index (HDI)

_____15. Human poverty index (HPI)

_____16. Command-and-control approach

_____17. Incentive-based approach

_____18. Green tax

_____19. Triple bottom line

j. A measurement index that combines three basic measures of human status: life expectancy; knowledge and education

k. A tax placed on environmentally harmful activities or emissions in an attempt to internalize some of the externalities that may be involved in the life cycle of those activities or products

l. The careful and responsible management and care for Earth and its resources

m. A strategy for pollution control that constructs financial and other incentives for lowering emissions based on profits and benefits

n. The U.S. organization that oversees all governmental efforts related to the environment, including science, research, assessment, and education

o. An agency of the U.S. Department of Labor, responsible for the enforcement of health and safety regulations

p. A worldview that holds that humans are just one of many species on Earth, all of which have equal intrinsic value

q. A strategy for pollution control that involves regulations and enforcement mechanisms

r. A global institution that provides technical and financial assistance to developing countries with the objectives of reducing poverty and promoting growth, especially in the poorest countries

s. A measurement index developed by the United Nations to investigate the proportion of a population suffering from deprivation in a country with a high HDI

Chapter 20 Review Exercises

Check Your Understanding

Review "Learning Objectives Revisited" on page 723 of your textbook. Compare the notes you took while reading each module. Complete these exercises to review the chapter.

1. Explain the difference between supply and demand.

2. What are the eight Millennium Development Goals?

3. What was Dr. Wangari Maathai's major contribution to the field of environmental science?

Practice for Free-Response Questions

Complete this exercise to build and practice the skills you will need to answer free-response questions on the exam. Use a separate sheet of paper if necessary.

What is GDP? Describe how an increase in a nation's GDP per capita can benefit the environment.

Review and Reflect

Complete these activities to solidify your knowledge of the chapter concepts and key terms. Use a notebook or a separate sheet of paper if necessary.

1. Review your key terms table for each module.

    (a) Which words did you already know? Which were new to you?
    (b) Write a new sentence using each key term.
    (c) Create a set of flash cards that includes each key term. Use the cards to review terms that were new or challenging.
    (d) When you feel comfortable with the new or challenging terms, review all of the cards, including those with familiar terms.
    (e) Save your cards to review before an exam.

2. What are the challenging concepts from this chapter?

    (a) Identify any concepts you found particularly challenging in this chapter.
    (b) Create a list of topics you need to review in preparation for an exam.

3. What questions do you have about concepts in the chapter?

    (a) Note any further questions you might have about material in the chapter.
    (b) Work with a partner to discuss these questions and ask your teacher for help as needed.

4. Write five possible multiple-choice questions based on this chapter. Work with a partner to quiz each other in preparation for an exam.

Unit 8 Multiple-Choice Review Exam

Choose the best answer.

1. Which greenhouse gas traps the majority of outgoing infrared radiation?
    (A) Methane
    (B) Water vapor
    (C) Carbon dioxide
    (D) Sulfur dioxide
    (E) Nitrogen dioxide

2. Which does NOT cause a decline in biodiversity?
    (A) Habitat loss
    (B) Alien species
    (C) Overharvesting
    (D) Pollution
    (E) Banning lead in gasoline

3. The greatest cause of biodiversity decline is
    (A) habitat loss.
    (B) alien species.
    (C) overharvesting.
    (D) pollution.
    (E) banning lead in gasoline

4. Which does NOT describe invasive species?
    (A) They spread rapidly.
    (B) They have no natural enemies.
    (C) They cause harmful effects on native species.
    (D) They are not typically introduced by humans.
    (E) They can outcompete native species.

5. Based on those species for which scientist have reliable data, which is most at risk of becoming threatened or near-threatened with extinction?
    (A) Birds
    (B) Mammals
    (C) Amphibians
    (D) Reptiles
    (E) Fish

6. The greenhouse effect happens when
    (A) the ozone layer is decreased.
    (B) infrared radiation is absorbed and emitted back to Earth.
    (C) UV light is trapped at Earth's surface.
    (D) greenhouse gasses emit UV light.
    (E) CFCs destroy the stratospheric ozone layer.

7. Which has the greatest greenhouse warming potential?
    (A) Water vapor
    (B) Carbon dioxide
    (C) Chlorofluorocarbons
    (D) Methane
    (E) Nitrous oxide

8. Which greenhouse gas comes from automobiles?
    (A) Nitrous oxide
    (B) Methane
    (C) Sulfur dioxide
    (D) Ozone
    (E) Chlorofluorocarbons

9. Which is an anthropogenic cause of greenhouse gas release?
        I. Volcanic eruptions
        II. Decomposition
        III. Coal burning power plants
    (A) I only
    (B) II only
    (C) III only
    (D) I and II
    (E) I, II and III

10. Which is NOT an anthropogenic source of greenhouse gas release?
    (A) Burning fossil fuels
    (B) Agriculture
    (C) Deforestation
    (D) Landfills
    (E) Use of asbestos

11. Which produces greenhouse gases and can also increase levels of mercury in the environment?
    (A) Refrigerators
    (B) Coal
    (C) Automobiles
    (D) Landfills
    (E) Agricultural practices

12. Which does NOT produce methane?
    (A) Livestock
    (B) Sewage treatment plants
    (C) Cement manufacturing
    (D) Wetlands
    (E) Termites

13. Which country produces the most carbon emissions?
    (A) The United States
    (B) China
    (C) Australia
    (D) India
    (E) England

14. Why do levels of carbon dioxide in Earth's atmosphere vary seasonally?
    (A) More fossil fuels are burned in the winter for heat.
    (B) More fossil fuels are burned in the summer for travel.
    (C) Livestock production increases each spring.
    (D) Photosynthesis varies each season.
    (E) Landfills increase in size as more trash is produced during summer months.

Use the following graph to answer question 15.

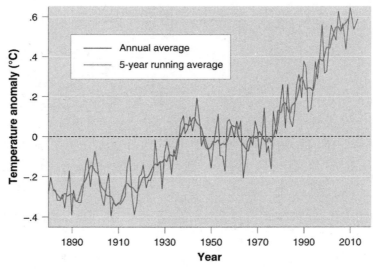

Figure 63.3
Environmental Science for AP®, Second Edition
Data from http://data .giss.nasa.gov/gistemp/graphs_v3/Fig.A2.gif

15. According to the graph
    (A) temperature has been increasing steadily since 1975.
    (B) from 1890 to 2010, average temperatures over 5 year periods do not fluctuate.
    (C) the highest temperature recorded was in 1940.
    (D) temperature has been declining recently.
    (E) the coldest temperature recorded was in 1920.

16. Which in NOT a likely effect of warmer temperatures on the environment?
    (A) Melting polar ice caps
    (B) Rising sea levels
    (C) Heat waves
    (D) Increased storms
    (E) Larger hole in the stratospheric ozone layer.

17. Which agreement aims to control the emissions that contribute to global warming?
    (A) Montreal Protocol
    (B) Kyoto Protocol
    (C) Convention on Biological Diversity
    (D) Clean Air Act
    (E) Climate Change Act

18. _____refers to storing carbon in agricultural soils to return atmospheric carbon to longer-term storage in the form of plant biomass.
    (A) Agricultural restoration
    (B) Soil reclamation
    (C) Carbon sequestration
    (D) Carbon dioxide scrubber
    (E) Greenhouse gas depletion

Use the following graph to answer question 19.

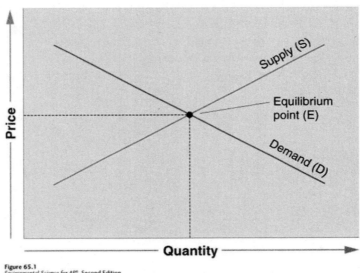

Figure 65.1
Environmental Science for AP, Second Edition
© 2015 W.H. Freeman and Company

19. In the graph, what does the equilibrium point represent?
    (A) Where price exceeds quantity supplied
    (B) Where quantity supplied exceeds price
    (C) Where quantity supplied exceeds quantity demanded
    (D) Where quantity supplied and quantity demanded are equal
    (E) Where quantity demanded exceeds quantity supplied

20. Which location in a biosphere reserve has the most human activity?
    (A) Buffer zone
    (B) Transition area
    (C) Core zone
    (D) Research zone
    (E) Settlement area

21. What does the Kuznets curve illustrate?
    (A) As per capita income decreases, environmental degradation decreases.
    (B) As per capita income decreases, environmental degradation increases.
    (C) As per capita income increases, environmental degradation decreases.
    (D) As per capita income increases, environmental degradation increases.
    (E) As per capita income increases, environmental degradation first increases and then decreases.

22. The view that nature has an instrumental value to provide for human needs reflects
    (A) an anthropocentric worldview.
    (B) a biocentric worldview.
    (C) an ecocentric worldview.
    (D) the precautionary principle.
    (E) a positive externality.

23. Taxes on environmentally harmful activities are known as
    (A) externalities.
    (B) cost/benefit taxes.
    (C) green taxes.
    (D) luxury taxes.
    (E) cradle to grave taxes.

Full-Length Practice Exam 1

This full-length practice exam contains two parts. Part I consists of 100 multiple-choice questions and Part II consists of four free-response questions.

You will have 90 minutes to complete the multiple-choice section of the exam. This section counts for 60 percent of the exam grade. As you will not be penalized for incorrect answers, you should answer every question on the test. If you do not know an answer to a question, try to eliminate any incorrect answer choices and take your best guess. Do not spend too much time on any one question. If you know the question is going to take a while to solve, you should skip it and come back to it at the end.

You will have 90 minutes to complete the free-response section of the exam. This section counts for 40 percent of the overall exam grade. Be sure to answer each part of the question and to provide thorough explanations using the terms and themes you have learned in the course. Also be sure to show your work whenever you use math to solve a problem.

Calculators are not allowed on any portion of the exam.

SECTION I: Multiple-Choice

Chose the best answer for questions 1-100.

1. In a country going through a demographic transition and becoming more industrialized
   (A) birth rates rise.
   (B) death rates rise.
   (C) birth rates fall.
   (D) death rates remain high.
   (E) death rates rise as birth rates fall.

2. Which is NOT associated with passive solar energy?
   (A) Putting a photovoltaic cell on the roof
   (B) Putting blinds on a window
   (C) Planting a deciduous tree outside a west facing window
   (D) Installing a living roof
   (E) Adding extra insulation in the walls and attic of a house

3. Which is the correct order of soil particles from largest to smallest?
   (A) Sand, clay, silt
   (B) Silt, clay, sand
   (C) Sand, silt, clay
   (D) Clay, silt, sand
   (E) Silt, sand, clay

4. A scientist does an experiment with brine shrimp and a particular pesticide and finds that 10 ml of the pesticide kills half of the shrimp. What has the scientist found?
   (A) The LD-50
   (B) The effective dose
   (C) The toxicity level
   (D) The ED-50
   (E) The teratogen level

5. Which is an example of an indicator species?
   (A) Elephant
   (B) Oak tree
   (C) Bald Eagle
   (D) Leopard frog
   (E) Box turtle

6. Radioactive waste generated in the United States is currently being stored
   (A) At Yucca Mountain.
   (B) in the ocean.
   (C) in Mexico.
   (D) at the nuclear plants where the waste is produced.
   (E) in designated landfills.

7. Which is a concern about excessive use of fertilizer in yards and gardens?
   (A) It will lead to an increased numbers of pests.
   (B) It will contribute to eutrophication in nearby waterways.
   (C) It will hasten the extinction of endangered species.
   (D) It will damage the niche of a keystone species.
   (E) It will lead to a drastic pH drop in lakes.

8. Approximately 50 percent of all coal reserves worldwide are found in which countries?
   (A) China, Japan, India
   (B) United States, Canada, Mexico
   (C) China, United States, Russia
   (D) Russia, China, Mexico
   (E) India, Brazil, Canada

9. Which is NOT a greenhouse gas?
   (A) Carbon dioxide
   (B) Methane
   (C) CFCs
   (D) Sulfur dioxide
   (E) Water vapor

10. Electrostatic precipitators and wet scrubbers are used to remove particulates and SO_2 from
    (A) cars.
    (B) industrial plants.
    (C) sewage treatment plants.
    (D) landfills.
    (E) contaminated soil.

11. Alum, which is used in sewage treatments plants as a flocculating reagent, helps
    (A) screen out the sewage.
    (B) decontaminate the sewage.
    (C) get rid of aerobic microorganisms.
    (D) the sewage to clump and sink.
    (E) aerate the sewage.

12. CFCs in the stratosphere break down ozone molecules. Why was this a concern?
    (A) Ozone is an important component of oxygen.
    (B) Ozone helps keep Earth's temperature stable.
    (C) Ozone in the stratosphere is a respiratory irritant.
    (D) Ozone protects Earth's surface from damaging UV light.
    (E) Ozone helps remove the carbon dioxide in our atmosphere.

13. Why might a country have a replacement level fertility as high as two children per person?
    (A) Women are working outside the home.
    (B) The life span is only 45 years.
    (C) The country has a high infant mortality rate.
    (D) Children are not seen as a necessity.
    (E) The country has a large population of women above childbearing age.

14. The efficiency of a typical coal-fired electricity plant is
    (A) 100 percent.
    (B) 75 percent.
    (C) 50 percent.
    (D) 30 percent.
    (E) 10 percent.

15. Habitat loss, invasive species, pollution, overpopulation, climate change, and overharvesting have contributed to
    (A) increasing the number of endangered species.
    (B) mountain top removal of coal.
    (C) cultural eutrophication in our water ways.
    (D) ozone depletion.
    (E) the amount of phosphorus in the atmosphere.

16. Arable land on Earth is
    (A) decreasing.
    (B) increasing.
    (C) being converted into nature preserves.
    (D) being developed for housing.
    (E) being converted into forest.

17. A city with a population of 10,000 has 50 births, 30 deaths, 30 immigrants and 10 emigrants in a year. What is the net annual percentage growth rate?
    (A) 120 percent
    (B) 40 percent
    (C) 8 percent
    (D) 0.4 percent
    (E) 10 percent

18. The Earth is tilted approximately 23 degrees. This causes
    (A) the seasons.
    (B) the tides.
    (C) the solar lights.
    (D) the Coriolis effect.
    (E) hurricanes.

19. If the current human population is approximately 6.8 billion and is growing at an annual rate of 1.17 percent, approximately how many births will occur next year at this rate?
    (A) 8×10^5
    (B) 8×10^6
    (C) 8×10^7
    (D) 8×10^8
    (E) 8×10^9

20. Radon is found in bedrock and can seep into homes through basements and other means. Why is radon a concern?
    (A) Radon causes asthma.
    (B) Radon causes brain damage in children.
    (C) Radon damages foundations.
    (D) Radon causes lung cancer.
    (E) Radon causes cataracts.

21. Which is NOT a disadvantage of using nuclear power?
    (A) Materials could be subject to a terrorist threat.
    (B) Storage solutions are difficult to find.
    (C) People have safety concerns regarding radiation exposure.
    (D) Nuclear power plants generate significant CO_2 emissions.
    (E) Natural disasters such as earthquakes can damage nuclear facilities.

22. Which of the following is NOT an advantage of genetically modified crops?
    (A) They are pest resistant.
    (B) They need less water.
    (C) They are more nutritious.
    (D) They have a greater yield.
    (E) They are less expensive.

23. Which is a nonrenewable energy source?
    (A) Nuclear
    (B) Solar
    (C) Wind
    (D) Geothermal
    (E) Biomass

24. Which is NOT typically included in the price of a good or service?
    (A) The cost of materials
    (B) The cost of labor
    (C) The cost of externalities
    (D) The cost of transportation
    (E) Profits

25. Weather on Earth is found in which layer of the atmosphere?
    (A) Troposphere
    (B) Stratosphere
    (C) Mesosphere
    (D) Thermosphere
    (E) Exosphere

Use the following graph to answer questions 26 and 27.

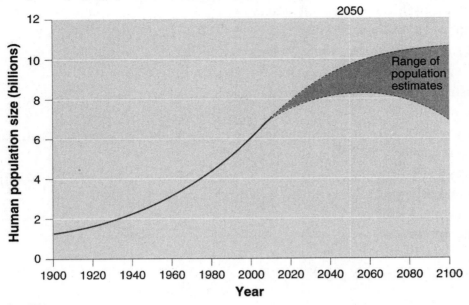

Figure 22.4
Environmental Science for AP®, Second Edition
After Millennium Ecosystem Assessment, 2005

Projected World Population Growth

26. According to the graph, from 1800 to 2010 human population displayed
    (A) linear growth.
    (B) bell-shaped growth.
    (C) exponential growth.
    (D) a sharp decline.
    (E) a slow decline.

27. If the size of the human population continues to grow at the high estimate, what will be the approximate population of Earth in the year 2300?
    (A) 11 billion
    (B) 8 billion
    (C) 6 billion
    (D) 20 billion
    (E) 14 billion

28. Consider two different communities each with 100 organisms.
   - Community 1 has 10 different species with 10 of each species.
   - Community 2 has 5 different species with the following numbers:
     - species 1 = 40
     - species 2 = 20
     - species 3 = 20
     - species 4 = 10
     - species 5 = 10

What can we tell about community 1 compared to community 2?
(A) Community 1 has more species richness than community 2.
(B) Community 2 has more species richness than community 1.
(C) Community 2 has more species evenness than community 1.
(D) Community 1 has been more disturbed than community 2.
(E) Community 1 is more stable than community 2.

Use the following diagram to answer questions 29 and 30.

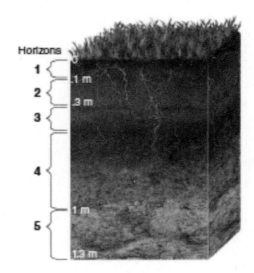

29. Which is a layer of organic detritus?
   (A) O
   (B) A
   (C) B
   (D) C
   (E) E

30. Which layer is known as the subsoil and has very little organic matter?
    (A) O
    (B) A
    (C) B
    (D) C
    (E) E

31. If a population is growing at a rate of 7 percent per year, how many years will it take for the population to double?
    (A) 5 years
    (B) 10 years
    (C) 17 years
    (D) 25 years
    (E) 70 years

32. When is the level of ozone in the Northern Hemisphere at its lowest?
    (A) January
    (B) March
    (C) July
    (D) September
    (E) December

33. Where is the world's largest hydroelectric dam?
    (A) United States
    (B) Brazil
    (C) Canada
    (D) China
    (E) India

34. Which causes most cultural eutrophication?
    (A) Carbon
    (B) Sulfur
    (C) Oxygen
    (D) Nitrogen
    (E) Calcium

35. A scientist discovers fecal coliform in a river. Which is the most likely source?
   (A) Chemical fertilizers used by nearby homeowners
   (B) Release of an invasive species
   (C) A faulty sewage treatment plant
   (D) A factory dumping industrial waste
   (E) Localized air pollution

36. What is the population density of a country that has 9.6 million km^2 of land area and 1,331 million people?
   (A) 138.6 people per km^2
   (B) 0.007 people per km^2
   (C) 12,777 people per km^2
   (D) 1,321.4 people per km^2
   (E) 1,331 people per km^2

37. Which is a consequence of increased greenhouse gas concentrations in Earth's atmosphere?
   (A) Too much UV light reaches the troposphere.
   (B) IR radiation becomes trapped in the troposphere.
   (C) Too much UV light reaches the stratosphere.
   (D) IR radiation becomes trapped in the stratosphere.
   (E) The troposphere releases too much IR radiation back into space.

38. A niche specialist
   (A) is found at a high trophic level.
   (B) has a highly varied diet.
   (C) can colonize new areas rapidly.
   (D) is engaged in mutualism.
   (E) is especially susceptible to environmental change.

39. Human population growth has increased dramatically in the last 100 years because of
   (A) modern medicine.
   (B) genetically modified crops.
   (C) access to reliable birth control.
   (D) climate change.
   (E) immigration.

40. Erosion is least likely to be caused by
   (A) sustainable agriculture.
   (B) overgrazing.
   (C) construction projects.
   (D) logging.
   (E) deforestation.

41. The order of coal from the lowest energy content to highest energy content is
    (A) anthracite, lignite, sub-bituminous, bituminous.
    (B) anthracite, sub-bituminous, bituminous, lignite.
    (C) sub-bituminous, bituminous, lignite, anthracite.
    (D) lignite, sub-bituminous, bituminous, anthracite.
    (E) sub-bituminous, bituminous, anthracite, lignite.

42. Which fossil fuel has the largest, most accessible supply on Earth?
    (A) Oil
    (B) Natural gas
    (C) Nuclear
    (D) Biomass
    (E) Coal

43. Which is NOT a concern about concentrated animal feeding operations (CAFOs)?
    (A) Adverse health effects of high density populations
    (B) An increase in antibiotic resistant strains of microorganisms
    (C) The amount of land they use
    (D) Groundwater contamination from animal waste
    (E) Waterway pollution from animal waste runoff

44. No-till farming leads to
    (A) a decrease in water erosion.
    (B) an increase in wind erosion.
    (C) an increase in water erosion.
    (D) overfertilization.
    (E) an increase in microorganisms.

45. Which heavy metal is produced by burning coal?
    (A) Lead
    (B) Mercury
    (C) Arsenic
    (D) Iron
    (E) Titanium

46. Which type of irrigation is 95 percent efficient?
    (A) Spray irrigation
    (B) Drip irrigation
    (C) Flood irrigation
    (D) Pivot irrigation
    (E) Furrow irrigation

47. Economic activity that creates a cost not borne by either the buyer or the seller is
    (A) a negative externality.
    (B) a positive externality.
    (C) an internalized cost.
    (D) a tragedy of the commons.
    (E) a fixed cost.

48. Your air conditioner uses 600 watts of electricity per hour on a daily basis, and your energy cost is $.10 per kWh. Assuming there are 30 days in a month, how much does the electricity used by the air conditioner cost you per month?
    (A) $1.30
    (B) $4.32
    (C) $43.20
    (D) $13.01
    (E) $1.80

49. If approximately 1.1 billion people in the world do not have access to clean water, approximately what percent of the world does not have clean water?
    (A) 10 percent
    (B) 15 percent
    (C) 25 percent
    (D) 40 percent
    (E) 50 percent

50. If you have 1,500 cows, and each cow produces 50 liters of manure a day, how many liters of manure would be produced in 45 days?
    (A) 33 kL
    (B) 337 kL
    (C) 3,375 kL
    (D) 33,750 kL
    (E) 337,500 kL

51. Which layer of the atmosphere is the most dense?
    (A) Troposphere
    (B) Stratosphere
    (C) Mesosphere
    (D) Thermosphere
    (E) Exosphere

52. Which biome has permafrost?
    (A) Tundra
    (B) Grassland
    (C) Tropical rainforest
    (D) Temperate rainforest
    (E) Desert

53. Which scientist is credited with the theory of evolution by natural selection?
    (A) Watson
    (B) Crick
    (C) Mendeleev
    (D) Muir
    (E) Darwin

54. After several years of using a particular pesticide, a farmer notices that it is killing fewer insect pests. This is probably because
    (A) the insect pests have become resistant to the pesticide.
    (B) the pesticide has leeched essential nutrients from the field.
    (C) other conditions have become more favorable for the insect pests.
    (D) the insects have migrated to other cropland.
    (E) eutrophication in local waterways has harmed insect breeding grounds.

55. What are the negative environmental problems associated with using biomass for energy?
    I. Deforestation
    II. Soil erosion
    III. Air pollution
    (A) I only
    (B) II only
    (C) III only
    (D) I and II
    (E) I, II, and III

Match the energy source in numbers 56-60 with the characteristic that describes it.

56. ____ Solar	(A) does not come from the sun
57. ____ Wind	(B) is used to cook with in many developing nations
58. ____ Geothermal	(C) is growing in popularity in Europe
59. ____ Biomass	(D) uses photovoltaic cells
60. ____ Hydroelectric	(E) uses falling water to make energy

Use the food chain below to answer question 61.

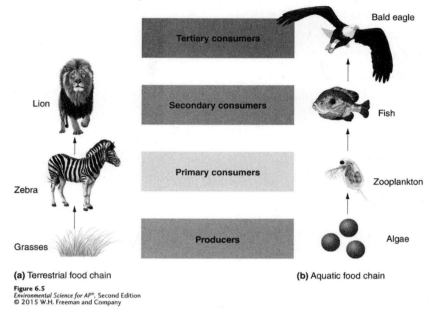

(a) Terrestrial food chain   **(b)** Aquatic food chain

Figure 6.5
Environmental Science for AP®, Second Edition
© 2015 W.H. Freeman and Company

61. If there are 10,000 kilocalories of energy available at the producer level, how many kilocalories will be available at the tertiary level?
    (A) 10,000
    (B) 1,000
    (C) 100
    (D) 10
    (E) 9

62. "Solar energy + 6H$_2$O + 6 CO$_2$ → C$_6$H12O$_6$ + 6O$_2$" is the formula for
    (A) respiration.
    (B) photosynthesis.
    (C) chemosynthesis.
    (D) cellular respiration.
    (E) primary productivity.

63. A forest ecosystem has an NPP of 3.05 kg C/m^2/year and a GPP of 4.5 kg C/m^2/year. How much carbon is lost to respiration?
    (A) 1.45 kg C/m^2/year
    (B) 7.55 kg/C/m^2/year
    (C) -1.45 kg C/m^2/year
    (D) 13.73 kg C/m^2/year
    (E) 17.27 kg C/m^2/year

64. When air rises, its
    (A) pressure increases and it expands.
    (B) pressure decreases and it expands.
    (C) pressure increases and it contracts.
    (D) pressure decreases and it contracts.
    (E) pressure and volume remain the same.

65. Which is NOT a component of integrated pest management?
    (A) Predator bugs
    (B) Pheromones
    (C) Crop rotation
    (D) Chemical pesticides
    (E) Irradiated insects

66. Which environment is most susceptible to desertification?
    (A) Rainforests
    (B) Grasslands
    (C) Areas near other deserts
    (D) Tundra
    (E) Taiga

67. In the United States, most MSW is
    (A) burned.
    (B) sent to landfills.
    (C) recycled.
    (D) composted.
    (E) shipped to other countries.

Use the graph below to answer questions 68 and 69.

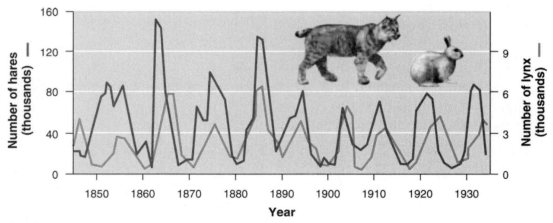

Figure 19.5
Environmental Science for AP®, Second Edition
Data from Hudson's Bay Company

68. According to the graph, when there are more hares than lynx,
    (A) the number of hares decrease.
    (B) the number of lynx grow quickly.
    (C) the lynx population grows slowly.
    (D) the hare population remains stable.
    (E) the lynx population remains stable.

69. The graph shows
    (A) carrying capacity.
    (B) survivorship curves.
    (C) population regulations.
    (D) predator-prey relationships.
    (E) exponential growth.

70. Without the law of thermodynamics, efficiency would be
    (A) 0 percent.
    (B) 10 percent.
    (C) 50 percent.
    (D) 75 percent.
    (E) 100 percent.

71. Which ecosystem exhibits the most biodiversity?
    (A) Coral reef
    (B) Deciduous forest
    (C) Desert
    (D) Tundra
    (E) Taiga

Use the following diagram to answer questions 72 and 73.

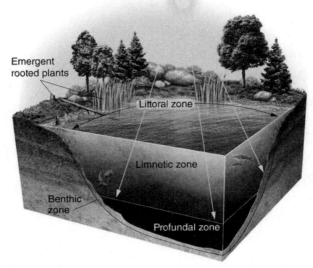

Figure 13.3
Environmental Science for AP®, Second Edition
© 2015 W.H. Freeman and Company

72. In which zone of the lake would you expect to find the least amount of photosynthetic algae?
    (A) Littoral
    (B) Limnetic
    (C) Profundal
    (D) Benthic
    (E) Euphotic

73. In which zone of the lake would you find algae but no rooted plants?
    (A) Littoral
    (B) Limnetic
    (C) Profundal
    (D) Benthic
    (E) Euphotic

74. How might deforestation affect a river system?
    (A) Increase sediment levels
    (B) Increase oxygen levels
    (C) Increase the presence of invasive species
    (D) Increase the number of fish present
    (E) Lower the average temperature of the water

75. Which is likely to occur with geographic isolation?
    (A) Sympatric speciation
    (B) Extinction
    (C) Allopatric speciation
    (D) Polyploidy
    (E) Evolution by artificial selection

76. The theory of plate tectonics describes how
    (A) animals adapt to changes in their environment.
    (B) Earth's lithosphere is divided into moving sections.
    (C) plants on Earth evolved from a single species.
    (D) molten material from Earth's mantle rises to the lithosphere.
    (E) how volcanoes contribute to air pollution problems.

77. An earthquake that measures 9 on the Richter Scale is how many times stronger than an earthquake that measures 7 on the Richter Scale?
    (A) 1
    (B) 10
    (C) 100
    (D) 1000
    (E) 10,000

78. Igneous rocks come from
    (A) magma.
    (B) other rocks.
    (C) sediment.
    (D) erosion.
    (E) weathering.

Use the following diagram to answer questions 79 and 80.

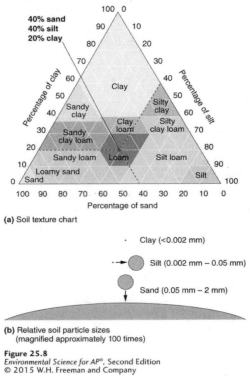

(a) Soil texture chart

Clay (<0.002 mm)

Silt (0.002 mm – 0.05 mm)

Sand (0.05 mm – 2 mm)

(b) Relative soil particle sizes
(magnified approximately 100 times)

Figure 25.8
Environmental Science for AP®, Second Edition
© 2015 W.H. Freeman and Company

79. The perfect mixture of sand, silt and clay is known as
    (A) silty clay.
    (B) sandy clay loam.
    (C) loam.
    (D) sand.
    (E) clay.

80. You have a soil sample of 20 percent sand, 30 percent silt, 50 percent clay. What kind of soil is it?
    (A) Sand
    (B) Silt
    (C) Clay
    (D) Loam
    (E) Silty clay loam

81. Leachate is the name for
    (A) waste products from invertebrates that live in the soil.
    (B) the toxic residue that remains after incinerating MSW.
    (C) liquid that contains pollutants from MSW or contaminated soil.
    (D) radioactive waste from a nuclear plant.
    (E) soil that erodes into into a waterway after deforestation.

82. Which of the following takes the most space in a typical landfill?
    (A) Compost
    (B) Aluminum
    (C) Plastic
    (D) Glass
    (E) Paper

83. An environmental disaster has occurred in the Aral Sea because
    (A) toxic waste was dumped in the water.
    (B) rivers were diverted and salinity increased.
    (C) a nuclear power plant released harmful levels of radiation.
    (D) An earthquake damaged surrounding infrastructure.
    (E) a tsunami destroyed the coastline and killed many people.

84. Farmers can use biological controls to reduce pesticides used on crops. However, sometimes biological controls can
    (A) lead to the evolution of a new species.
    (B) overpopulate an area.
    (C) cause the pest to multiply.
    (D) become resistant.
    (E) eat both beneficial and pest species.

85. Many farmers have found that over a number of years they must spray more and more pesticide on their crops for the same result. This is known as
    (A) insect migration.
    (B) the pesticide treadmill.
    (C) GMOs.
    (D) speciation.
    (E) the founder effect.

86. What law enforces the cleanup of hazardous waste sites?
    (A) CERCLA (Superfund)
    (B) FIFRA
    (C) RCRA
    (D) Clean Water Act
    (E) Clean Air Act

87. Which would be the best example of a keystone species?
    (A) Cockroach
    (B) Spider
    (C) Frog
    (D) Bald eagle
    (E) Elephant

88. Which chemical is responsible for acid deposition?
    (A) Carbon dioxide
    (B) Mercury
    (C) Sulfur Dioxide
    (D) CFCs
    (E) Lead

89. Which is a characteristic of an *r*-selected species?
    (A) Small size
    (B) Specialist
    (C) Often hunted
    (D) At risk for endangerment
    (E) Few offspring

90. Sulfur dioxide is a component of industrial smog and the precursor to acid deposition. Which pollution control measure is most effective for sulfur dioxide?
    (A) Catalytic converter
    (B) Scrubber
    (C) Electrostatic precipitator
    (D) Bag house filter
    (E) Bioremediation

91. Which is NOT associated with sediment pollution?
    (A) Inability of fish and benthic organisms to obtain enough oxygen
    (B) Reduction in productivity of plants.
    (C) Increase in nutrient pollution
    (D) Increased construction and agriculture
    (E) Loss of aquatic life from thermal shock

92. Because the U.S. population has a TFR of 1.9 and high net migration, it may be best described as
    (A) a country experiencing population momentum.
    (B) a country with declining population growth.
    (C) a country with a large aging population.
    (D) a country with stable population growth.
    (E) a country with rapid population growth.

93. Which international agreement effectively banned the use of CFCs?
    (A) Montreal Protocol
    (B) Kyoto Protocol
    (C) CITIES
    (D) REACH
    (E) Millennium Development Goals

94. Which is the best example of a positive externality?
    (A) An apple farmer provides a beekeeper with a source of nectar for honey.
    (B) A beaver creates a natural dam providing habitats for many other species.
    (C) A wolf maintains population levels of herbivores.
    (D) Chemical pesticides used by a farmer flow into local water supplies.
    (E) A wetland provides flood control for surrounding communities.

95. Which is NOT used to remediate oil spills?
    (A) Floating booms
    (B) Genetically engineered bacteria
    (C) Drift nets
    (D) Chemical dispersal agents
    (E) Absorbent materials

96. Which would be an example of a pollution remediation technique?
    (A) Removing catalytic converters from cars
    (B) Installing wet scrubbers on power plants
    (C) Burning biomass for energy
    (D) Directing waste water effluent into a stream
    (E) Returning leachate to the environment

97. Which is the doubling time of the United States if the growth rate is 0.7 percent?
    (A) 10 years
    (B) 0.01 years
    (C) 70 years
    (D) 100 years
    (E) 700 years

98. The Kyoto Protocol addresses which environmental concern?
    (A) Ozone depletion
    (B) Habitat loss
    (C) Biodiversity loss
    (D) Marine mammals being killed
    (E) Climate change

99. Which is a negative externality of producing electricity?
    (A) The cost of the fuel
    (B) The cost of damages caused by air pollution
    (C) The price consumers pay for their electricity
    (D) A tax on the electricity
    (E) The cost of maintaining the plant that generates electricity

100. Skin cancer rates have been increasing because of
    (A) climate change.
    (B) ozone depletion.
    (C) deforestation.
    (D) eutrophication.
    (E) deprivation.

SECTION II: Free-Response Questions

Write your answer to each part clearly. Support your answers with relevant information and examples. Where calculations are required, show your work.

1. Last night at city hall, the residents of New Hanover got into a heated debate over Genetically Modified Organisms (GMOs). Larry, a farmer, explained the need to use GMO wheat to make his land more productive and to produce more food. Mary, a concerned mother, stated that there is no way to convince her that GMOs are safe for her children and she will not serve them in her home.

    (A) Describe two benefits of GMOs.

(B)   Describe how GMOs are created.

(C)   List and describe two possible reasons Mary is concerned about GMOs.

(D)   Larry's community would like to encourage more sustainable farming techniques. List and describe two sustainable farming methods he could consider incorporating on his farm.

2.   The Miller family has decided that due to the severe drought in their state, they are going to implement water conservation methods in their home. Currently, the family uses approximately 8,000 gallons of water a month for their family of four and they pay $3.45 per 100 cubic feet of water. There are 748 gallons in 100 cubic feet.

(A)   How much is the family's water bill per month?

(B)   The family installs water-saving shower fixtures in both bathrooms, which cuts the amount of water for showering in half. Each member of the family takes one shower a day and the average time spent in the shower is 10 minutes. The old shower fixture used 42 gallons every 10 minutes. How many gallons will be saved by the family per week with the water-saving fixtures?

(C) Other than the shower fixtures, the Miller family wants to install other water-saving devices in their home. List and describe TWO ways the family could conserve more water in their home.

(D) The family also wants to incorporate water-saving methods outside. Describe two ways they could accomplish this goal.

3. Scientists have been concerned about the effects of the depletion of the stratospheric ozone layer on the health of humans and ecosystems.

(A) Describe how the stratospheric ozone layer was damaged, including the chemicals that caused the damage.

(B) List the chemical reactions that took place in the stratosphere that caused the hole in the ozone layer.

(C) List and describe two environmental or human health issues related to the hole in the ozone layer.

(D) What is the name of the agreement that was signed in 1987 to take steps toward resolving the destruction of the ozone layer?

(E) Ozone has become a problem in the troposphere. List one human health issue related to this problem and explain what humans are doing to cause ozone to accumulate in the troposphere.

4. The graph below shows the phases of demographic transition.

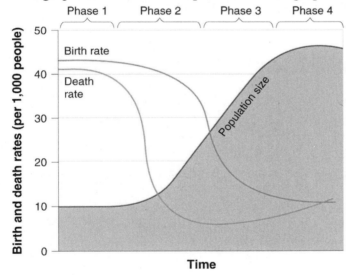

Figure 23.1
Environmental Science for AP®, Second Edition
© 2015 W.H. Freeman and Company

(A) Describe each phase of the demographic transition. Include a discussion of population size and living conditions.

(B) Give two reasons why a family in Phase 2 would have many children.

(C) If a country of 10,000 has 100 births, 80 deaths, 30 immigrants, and 20 emigrants in a typical year, what is the population growth rate?

(D) What is the doubling time for this country?

Full-Length Practice Exam 2

This full-length practice exam contains two parts. Part I consists of 100 multiple-choice questions and Part II consists of four free-response questions.

You will have 90 minutes to complete the multiple-choice section of the exam. This section counts for 60 percent of the exam grade. As you will not be penalized for incorrect answers, you should answer every question on the test. If you do not know an answer to a question, try to eliminate any incorrect answer choices and take your best guess. Do not spend too much time on any one question. If you know the question is going to take a while to solve, you should skip it and come back to it at the end.

You will have 90 minutes to complete the free-response section of the exam. This section counts for 40 percent of the overall exam grade. Be sure to answer each part of the question and to provide thorough explanations using the terms and themes you have learned in the course. Also be sure to show your work whenever you use math to solve a problem.
Calculators are not allowed on any portion of the exam.

SECTION I: Multiple-Choice
Choose the best answer for questions 1-100.

1. Pure water with a pH of 7 is how many times more basic than stomach fluid with a pH of 1?
   (A) 100
   (B) 1,000
   (C) 6
   (D) 6,000
   (E) 1,000,000

2. Which type of irrigation technique is only 75 percent efficient and involves digging trenches to fill with water?
   (A) Spray irrigation
   (B) Flood irrigation
   (C) Furrow irrigation
   (D) Drip irrigation
   (E) Hydroponic irrigation

3. If the average person in the United States uses 1,000 watts of electricity, 24 hours a day for 365 days per year, how many kW of energy does the average person use in a year?
   (A) 10 kW
   (B) 1000 kW
   (C) 3,650 kW
   (D) 3,650,000 kW
   (E) 8760 kW

Match the terms below with the correct definition.

4. _____ Accuracy

5. _____ Precision

6. _____ Uncertainty

(A) How close a measured value is to the actual or true value

(B) An estimate of how much a measured or calculated value differs from a true value

(C) How close the repeated measurements of a sample are to one another

(D) The number of times a measurement is replicated in data collection

(E) Data collection by taking repeated measurements

7. Which of the following is the correct equation for photosynthesis?
   (A) Energy + $6H_2O$ + 7 CO_2 → $C_6H_{12}O_6$ + $8O_2$
   (B) Energy + 6 H_2O + 6 CO_2 → $C_6H_{12}O_6$ + 6 O_2
   (C) Solar energy + 6 H_2O + 8 CO_2 → $C_6H_{12}O_6$ + 8 O_2
   (D) Solar energy + 8 H_2O + 8 CO_2 → $C_6H_{12}O_6$+ 12 O_2
   (E) Solar energy + 6 H_2O + 6 CO_2 → $C_6H_{12}O_6$ + 6 O_2

8. Which is NOT part of a nuclear reactor?
   (A) Containment structure
   (B) Steam generator
   (C) Control rods
   (D) Turbine
   (E) Scrubber

9. The estimate of the average number of children that each woman in a population will bear throughout her childbearing years is
   (A) the total fertility rate.
   (B) average life expectancy.
   (C) the crude birth rate.
   (D) family planning.
   (E) the infant mortality rate.

10. How much of the world's population does not have access to sufficient supplies of safe drinking water?
   (A) 1 out of every 100
   (B) 1 out of every 50
   (C) 1 out of every 25
   (D) 1 out of every 10
   (E) 1 out of every 6

11. Tectonic movement of Earth's plates is the result of
    (A) iron in Earth's core.
    (B) convection currents in the asthenosphere.
    (C) the density of the lithosphere.
    (D) the density of Earth's core.
    (E) subduction zones found near Japan and Argentina.

12. Which food production activity has the lowest energy subsidy?
    (A) Hunting and gathering
    (B) Raising grass-fed beef
    (C) Large feedlots to produce beef
    (D) Far-offshore fishing
    (E) Locally-produced food sold at a farmer's market

13. Rain that is slightly acidic can cause
    (A) chemical weathering.
    (B) physical weathering.
    (C) convection.
    (D) subduction.
    (E) eutrophication.

14. Which is NOT a part of the hydrologic cycle?
    (A) Transpiration
    (B) Condensation
    (C) Nitrification
    (D) Precipitation
    (E) Infiltration

15. Measured on the Richter scale, an earthquake with a magnitude of 7.0 is _____ times greater than an earthquake with a magnitude of 4.0.
    (A) 10
    (B) 100
    (C) 1,000
    (D) 10,000
    (E) 100,000

Match the steps in the sewage treatment process in numbers 16-18 with the correct description.

16. _____ Biological   (A) Solid waste settles out.

17. _____ Chemical   (B) Bacteria break down organic matter.

18. _____ Mechanical   (C) Chlorine, ozone, or ultraviolet lights are used.

19. A photovoltaic cell is used to
    (A) turn the sun's energy into electricity.
    (B) burn biomass fuel.
    (C) generate passive solar energy.
    (D) generate wind power.
    (E) generate electricity behind a dam.

20. The pesticide treadmill occurs when
    (A) bioaccumulation in predator species becomes pervasive.
    (B) a homeowner uses fertilizer frequently.
    (C) poor weather conditions make crops more vulnerable to pests.
    (D) pests begin to reproduce at a faster rate.
    (E) a farmer must switch pesticides because resistance develops.

21. Which would cause the greatest decline in biodiversity?
    (A) Selective cutting
    (B) Clear cutting
    (C) Shelter-wood harvesting
    (D) Seed-tree harvesting
    (E) No-till agriculture

22. Which is a disadvantage of tidal energy?
    (A) It is expensive to install.
    (B) It contributes to carbon dioxide emissions.
    (C) It is aesthetically displeasing.
    (D) It is available only near a coastline.
    (E) It pollutes the water.

23. Which is an environmental concern regarding the use of wind power?
    (A) The amount of air pollution generated
    (B) Resource depletion
    (C) Batteries stored in the windmill
    (D) Dependence on the amount of sunlight
    (E) Death of birds and bats that collide with turbines

24. Which is a biotic component of an ecosystem?
    (A) Soil
    (B) Sunlight
    (C) Nitrogen
    (D) Grass
    (E) Air

Match the terms in numbers 25-29 with the correct description.

25. _____ Nitrification
26. _____ Assimilation
27. _____ Mineralization
28. _____ Ammonification
29. _____ Nitrogen fixation

(A) Producers incorporate elements into their tissues.
(B) Decomposers break down organic matter found in dead bodies and waste and convert it into inorganic compounds.
(C) Ammonia is converted into nitrite and nitrate.
(D) Organisms convert nitrogen gas molecules directly into ammonia.
(E) Decomposers break down organic matter found in dead bodies and waste and convert it into inorganic ammonium.

30. A species that has a stronger influence on other species in its community than its abundance might suggest is known as
    (A) a prey species.
    (B) an indicator species.
    (C) a mutualistic species.
    (D) a pioneer species.
    (E) a keystone species.

31. Which is the least expensive, and most environmentally damaging method of harvesting trees?
    (A) Selective cutting
    (B) Clear cutting
    (C) Slash and burn
    (D) Logging
    (E) Strip cutting

Use the figure below to answer question 32.

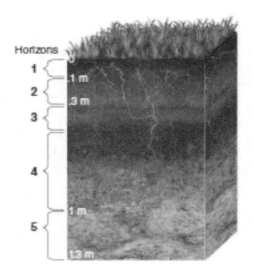

32. The soil layer that contains freshly fallen leaves is the
    (A) O horizon
    (B) A horizon
    (C) B horizon
    (D) C horizon
    (E) E horizon

33. Which describes a difference between the process of generating electricity using coal and the process of generating electricity using nuclear energy?
    (A) Nuclear energy generates steam and coal does not.
    (B) Coal produces air pollution and nuclear energy does not.
    (C) Coal waste must be stored for millions of years and waste from nuclear energy does not.
    (D) Coal-generated electricity is much more energy efficient than nuclear-generated electricity.
    (E) a generator is not needed in the production of electricity from nuclear energy but is needed in the in the production of electricity from coal.

34. Which chemical is most damaging to the stratospheric ozone layer?
    (A) Carbon
    (B) Fluorine
    (C) Methane
    (D) Sulfur dioxide
    (E) Carbon dioxide

35. Which is NOT an anthropogenic source of greenhouse gases?
    (A) Burning of oil
    (B) Agriculture
    (C) Logging
    (D) Sewage treatment plants
    (E) Volcanoes

36. Which does NOT produce methane?
    (A) termites.
    (B) landfills.
    (C) automobiles.
    (D) wetlands.
    (E) cattle farming.

Use the graph below to answer questions 37-39.

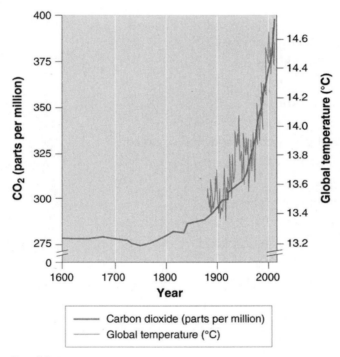

Figure 2.5
Environmental Science for AP®, Second Edition
Data from http://data.giss.nasa.gov/gistemp/graphs_v3/ and http://www .esrl.noaa.gov/gmd/ccgg/trends/#mlo_full

37. When carbon dioxide levels were at 300 ppm, what was the approximate global temperature in degrees Celsius?
    (A) 13.2
    (B) 14.1
    (C) 13.5
    (D) 13.8
    (E) 14.8

38. What is the approximate percent change in temperature from 1600 to 2000?
    (A) 11 percent increase
    (B) 25 percent increase
    (C) 32 percent increase
    (D) 45 percent increase
    (E) 76 percent increase

39. If recent trends in global surface temperatures continue, what year do you estimate global temperature will become 15.0 Celsius?
    (A) 2000
    (B) 2100
    (C) 2500
    (D) 3000
    (E) 3500

40. Monocropping can lead to
    (A) heavy flooding.
    (B) greater susceptibility of crops to pests.
    (C) introduction of invasive species.
    (D) uncontrollable fires.
    (E) significant erosion of topsoil.

41. Which is NOT a benefit of genetically modified organisms?
    (A) Increasing food production
    (B) Less loss of crops to pests
    (C) Reduced water requirements
    (D) Increased nutritional value
    (E) Spread of genetic material to wild plants

42. Which energy source is used the most in the United States?
    (A) Coal
    (B) Oil
    (C) Natural gas
    (D) Nuclear
    (E) Renewables

43. What type of rock is made from molten lava?
    (A) Igneous
    (B) Sedimentary
    (C) Metamorphic
    (D) Mineral
    (E) Marble

44. The net primary productivity of an ecosystem is 75 kg C/m^2/year, and producers require 20 kg C/m^2/year for their own respiration. What is the gross primary productivity of the ecosystem?
    (A) 10 kg C/m^2/year
    (B) 15 kg C/m^2/year
    (C) 50 kg C/m^2/year
    (D) 95 kg C/m^2/year
    (E) 30 kg C/m^2/year

45. Water is most dense at which temperature?
    (A) 0° Celsius
    (B) 32° Celsius
    (C) 100° Fahrenheit
    (D) 4° Celsius
    (E) 100° Celsius

46. The greenhouse effect occurs when
    (A) gases are trapped in the stratospheric ozone layer.
    (B) infrared radiation is absorbed by gases in Earth's atmosphere.
    (C) UV light is trapped at Earth's surface.
    (D) greenhouse gases absorb UV light.
    (E) CFCs destroy the stratospheric ozone layer.

47. Which is NOT characteristic of invasive species?
    (A) They are a threat to biodiversity.
    (B) They often do not have natural enemies.
    (C) They outcompete native species.
    (D) They remain concentrated in one area.
    (E) They can often reproduce quickly.

48. A scrubber on a coal burning power plant is designed to prevent the release of
    (A) sulfur dioxide.
    (B) nitrogen dioxide.
    (C) particulate matter.
    (D) carbon dioxide.
    (E) methane.

Use the following graph to answer question 49.

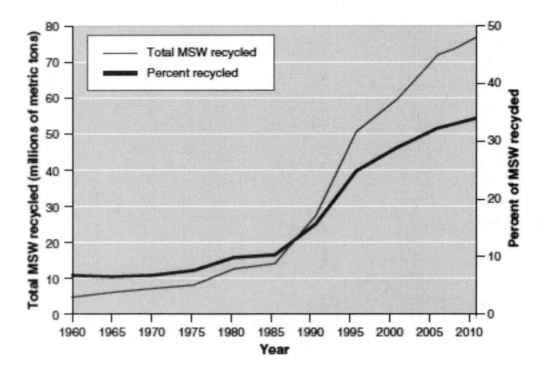

49. Use the graph above to calculate and compare the approximate percent change in recycled materials from 1970 to 1990.
    (A) 10 percent increase
    (B) 25 percent increase
    (C) 75 percent increase
    (D) 270 percent increase
    (E) 380 percent increase

50. In one year, a population of 10,000 has 200 births, 100 deaths, 60 immigrants and 30 emigrants. What is the population growth rate?
    (A) 1.3 percent
    (B) 9 percent
    (C) 90 percent
    (D) 2.4 percent
    (E) 24 percent

51. Cogeneration involves
    (A) using a fuel to generate electricity and heat.
    (B) using both coal and oil to create electricity.
    (C) increasing the capacity of nuclear power plants in major metropolitan areas.
    (D) substituting anthracite coal for low grade lignite coal.
    (E) operating power plants at 30 percent of maximum sustainable yield.

52. Freshwater is approximately what percentage of water on Earth?
   (A) 22 percent
   (B) 0.5 percent
   (C) 97 percent
   (D) 3 percent
   (E) 77 percent

53. 14 square miles is equal to _____ acres. (1 square mile = 640 acres)
   (A) 0.8960
   (B) 8.960
   (C) 89.60
   (D) 896.0
   (E) 8,960

54. The Coriolis effect is
   (A) the deflection of an object's path due to the rotation of Earth.
   (B) convection that creates air currents.
   (C) the transformation of arable land to desert because of global warming.
   (D) a change in the predator-prey relationship because of population cycles.
   (E) Increased crop yield due to planting crops parallel to the contour of the land.

55. If a population of 200 ducks increases to 500 ducks, what is the percent change in the population of ducks?
   (A) 1.5 percent increase
   (B) 15 percent increase
   (C) 150 percent increase
   (D) 50 percent increase
   (E) 25 percent increase

56. A country with a large population that lives in extreme poverty will
   (A) remain stable.
   (B) have a high infant mortality rate.
   (C) have a large environmental impact.
   (D) have a high GDP.
   (E) have a low emigration rate.

57. Which is NOT a natural source of greenhouse gases?
   (A) Volcanic eruptions
   (B) Decomposition
   (C) Digestion
   (D) Burning fossil fuel
   (E) Water vapor

58. Which is an example of a point source water pollutant?

(A) Agricultural lands

(B) Animal feedlots

(C) Runoff from parking lots

(D) Factory effluent

(E) Residential lawns

59. Which may affect global biodiversity?

    I. Loss of diversity at the genetic level

    II. A reduction in ecosystem services

    III. Increased habitat fragmentation

(A) I only

(B) II only

(C) III only

(D) I and II

(E) I and III

60. Which is an example of a *K*-selected species?

(A) Cockroach

(B) Fish

(C) Elephant

(D) Grasshopper

(E) Sea turtle

61. According to the World Health Organization (WHO) nearly half of the world's population is

(A) malnourished.

(B) anemic.

(C) obese.

(D) eating contaminated beef.

(E) suffering from malaria.

62. The El Niño-Southern Oscillation would bring what type of weather conditions to southern Africa and Southeast Asia?

(A) Warmer, drier

(B) Warmer, wetter

(C) Cooler, drier

(D) Unusually dry

(E) Unusually wet

63. Which occurs when one species becomes two in the absence of geographic isolation?
   (A) Reproductive isolation
   (B) Allopatric speciation
   (C) Sympatric speciation
   (D) Bottleneck effect
   (E) Founder effect

64. According to the theory of island biogeography, species richness increases as
       I.   island size increases.
       II. the distance of the island from mainland decreases.
       III. the distance from the mainland increases.
   (A) I and II
   (B) II and III
   (C) I and III
   (D) I only
   (E) I, II, and III

65. Which is a characteristic of a genetically modified organism (GMO)?
       I.   A GMO contains genetic material transferred from a different organism.
       II. A GMO can be engineered to be less palatable to pests.
       III. GMOs are used primarily in organic farming.
   (A) I only
   (B) I and II
   (C) II and III
   (D) I and III
   (E) III only

66. Which has encouraged population growth in suburban areas rather than urban areas?
       I. Increased reliance on automobiles
       II. Less expensive living costs in suburban areas
       III. Increased highway construction
   (A) I only
   (B) I and II
   (C) I and III
   (D) III only
   (E) I, II, and III

67. Which is the correct order from most moisture and least heat to least moisture and most heat?
    (A) peat, lignite, bituminous, anthracite
    (B) peat, bituminous, lignite, anthracite
    (C) anthracite, bituminous, lignite, peat
    (D) bituminous, anthracite, lignite, peat
    (E) bituminous, lignite, peat, anthracite

68. Which gas stays in the environment for up to 500 years and has the greatest global warming potential?
    (A) Water vapor
    (B) Carbon dioxide
    (C) Chlorofluorocarbons
    (D) Methane
    (E) Nitrous oxide

69. Which is an example of an anthropogenic activity?
    (A) Humans burning fossil fuels to generate electricity
    (B) Bees pollinating an apple tree
    (C) A volcanic eruption on a populated island
    (D) Water vapor rising from a lake
    (E) Coal deposits forming over millions of years

70. Which is NOT a location from which magma rises?
    (A) Divergent zones
    (B) Transform plate boundaries
    (C) Subduction zones
    (D) Hotspots
    (E) Points of seafloor spreading

71. By which process is igneous rock formed?
    (A) Weathering
    (B) Erosion
    (C) Convection
    (D) Compression
    (E) Cooling and crystallization

72. Which contributes to the greenhouse effect?
    I. Methane
    II. Water vapor
    III. Ozone
    (A) I
    (B) I and II
    (C) I, II, and III
    (D) I and III
    (E) II and III

73. Which is an accurate statement about wind energy?
    (A) Wind energy is potentially renewable.
    (B) Wind generated electricity is readily available in all areas.
    (C) Wind energy has no environmental disadvantages.
    (D) After turbine installation, there is little cost to harvest energy.
    (E) Wind energy contributes to a net increase in greenhouse gases.

74. When in the sewage treatment process is large debris filtered out by screens?
    (A) Primary treatment
    (B) Secondary treatment
    (C) Tertiary treatment
    (D) Both primary and secondary treatment
    (E) Both primary and tertiary treatment

75. A sample of radioactive waste has a half-life of 40 years and an activity level of 4 curies. After how many years will the activity level of this sample be 0.5 curies?
    (A) 40 years
    (B) 60 years
    (C) 80 years
    (D) 100 years
    (E) 120 years

76. What are the two reasons for the rapid growth of the human population over the past 8,000 years?
    (A) High infant mortality and treatment for HIV/AIDS
    (B) Medicine and technology
    (C) Smart growth and technology
    (D) Organic agriculture and improved communications
    (E) Lack of reliable birth control

77. Greenhouse gases in the atmosphere
    (A) prevent UV light from reaching Earth.
    (B) regulate temperatures near Earth's surface.
    (C) allow heat to be released back to space.
    (D) keep the ozone layer intact.
    (E) help Earth stay warm.

78. Which is NOT a top petroleum-producing county?
    (A) Saudi Arabia
    (B) Russia
    (C) the United States
    (D) Iran
    (E) Australia

79. Which is NOT an example of a secondary pollutant?
    (A) H_2SO_4
    (B) SO_3
    (C) O_3
    (D) H_2O_2
    (E) CO_2

80. Which statement about ozone depletion is NOT correct?
    (A) Ozone depletion is the result of automobile emissions and coal-fired electricity plants.
    (B) Ozone depletion occurs when CFCs are released into the atmosphere.
    (C) Ozone depletion occurs in the stratosphere.
    (D) Ozone depletion can lead to an increase in the incidence of skin cancer.
    (E) Ozone depletion allows more ultraviolet waves to pass through to the troposphere.

81. Which produces methane?
    (A) Aerosols
    (B) Air conditioners
    (C) Cement manufacturing
    (D) Wetlands
    (E) Automobiles

82. If a material has a radioactivity level of 100 curies and has a half-life of 50 years, how many half-lives will have occurred after 100 years?
    (A) 1
    (B) 2
    (C) 10
    (D) 1,000
    (E) 25

83. Thick curtains on windows can be an aspect of
    (A) active solar design.
    (B) photovoltaic systems.
    (C) energy star technology.
    (D) passive solar design.
    (E) a tiered rate system.

84. Soils found in tropical rain forests are
    (A) deep and nutrient rich.
    (B) rich in quartz sand.
    (C) highly porous.
    (D) quickly depleted of nutrients when the forest is removed.
    (E) highly permeable.

85. The current size of the human population is closest to
    (A) 300 million.
    (B) 300 billion.
    (C) 10 billion.
    (D) 1 trillion.
    (E) 7 billion.

86. The Ogallala aquifer is
    (A) located in the eastern United States.
    (B) a major source of water in California.
    (C) heavily polluted.
    (D) an unconfined aquifer.
    (E) being used unsustainably.

87. If a country's population growth rate is 7 percent, what is the country's doubling time?
    (A) 5 years
    (B) 35 years
    (C) 10 years
    (D) 42 years
    (E) 72 years

Use the following graph to answer question 88.

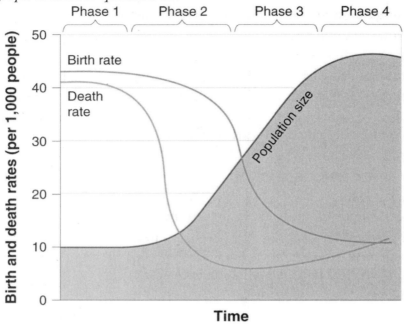

Figure 23.1
Environmental Science for AP®, Second Edition
© 2015 W.H. Freeman and Company

88. In the graph above, at which phase does the birth rate fall below the death rate?
   (A) Phase 1
   (B) Phase 1 and 2
   (C) Phase 3
   (D) Phase 4
   (E) Phase 1 and 4

89. In the graph above, at which phase is the population size relatively stable?
   (A) Phase 1
   (B) Phase 2
   (C) Phase 3
   (D) Phase 4
   (E) Phase 1 and 4

90. Which is not an infectious disease?
   (A) Meningitis
   (B) Ebola hemorrhagic fever
   (C) Tuberculosis
   (D) HIV/AIDS
   (E) Diabetes

91. Why do levels of carbon dioxide in the atmosphere vary with the seasons?
    (A) More fossil fuels are burned in the winter for heat.
    (B) More fossil fuels are burned in the summer for cooling.
    (C) Livestock production increases each spring.
    (D) Deciduous trees do not take in carbon dioxide in the fall and winter.
    (E) Convection currents in the air are more active in spring and summer.

92. Which is a major environmental impact associated with deforestation?
    (A) Removal of soil nutrients
    (B) Acid rain
    (C) Depletion of the stratospheric ozone layer
    (D) Increased municipal waste
    (E) Invasive species taking over the deforested area

93. Oil is generally found with
    (A) natural gas.
    (B) coal.
    (C) both natural gas and coal.
    (D) uranium.
    (E) heavy metals.

94. A lake that has low or no levels of nitrogen and phosphorous is
    (A) oligotrophic.
    (B) desalinated.
    (C) mesotrophic.
    (D) eutrophic.
    (E) acidic.

95. The process of recycling an aluminum can into a new aluminum can is an example of
    (A) closed-loop recycling.
    (B) the throw away society.
    (C) life-cycle analysis
    (D) the three R's.
    (E) open-loop recycling.

96. The process of desalinization
    (A) gives salt water a pH 7.5.
    (B) will help recharge the Ogallala Aquifer.
    (C) prevents saltwater intrusion.
    (D) causes saltwater intrusion.
    (E) is expensive and consumes energy.

97. Which laws was created in response to Love Canal?
    (A) RCRA
    (B) Brownfields
    (C) ESA
    (D) CERCLA
    (E) FIFRA

98. Fair distribution of Earth's resources is known as
    (A) community equality.
    (B) environmental equity.
    (C) empowerment.
    (D) the human development index.
    (E) stewardship.

99. Which describes the results of an experiment to determine the
    LD50 of a chemical?
    (A) 2 out of every 100 rats died
    (B) 25 out of every 50 rats died
    (C) 1 out of every 50 rats got sick
    (D) 50 out of every 100 rats got sick
    (E) 50 out of 1,000 rats died

100. According to the laws of thermodynamics
        I. energy is neither created nor destroyed.
        II. when energy is transformed, the quantity of energy remains the same.
        III. when energy is transformed, its ability to do work diminishes.
    (A) I only
    (B) II only
    (C) III only
    (D) I and III only
    (E) I, II, and III

SECTION II: Free-Response Questions

Write your answer to each part clearly. Support your answers with relevant information and examples. Where calculations are required, show your work.

1. The town of Maple Grove is concerned about water pollution because of sewage odor coming out of the local river. The town council has appointed a task force to investigate this problem.

   (A) Name 2 tests that could be performed to evaluate if sewage is contaminating the water. Describe how high levels of these contaminates would alter the water quality.

   (B) The citizens discover that their sewage treatment plant is not working properly. List the 2 stages of sewage treatment and describe the goal of each.

   (C) A member of the task force is concerned that a local livestock operation's feedlot may be contributing to the water pollution problem in the river. Propose a solution to address water pollution that occurs from a nearby animal feedlot.

   (D) The taskforce member explained that the livestock operation is a nonpoint source for water pollution. Explain why an animal feedlot is a nonpoint source. Explain the difference between point source and nonpoint source pollution.

2. A high school in Florida has decided to install photovoltaic panels on its roof. The cost to install the panels will be $15,000. Currently, the school pays $0.10 per kWh and the average monthly use for the school is 20,000 kWh. The panels will produce 2000 kWh per month.

(A) How much did the school pay during a 30-day month before the installation of the photovoltaic panels?

(B) How many years will it take for the school to recoup the cost of its purchase?

(C) If the school decides to invest $30,000 more in solar panels, how much more energy could they produce? (Express your answer in kWh.)

(D) Describe 2 practices the school could implement to lower its energy use.

(E) The school decides to incorporate passive solar design as a way to decrease its energy consumption. Name and describe 2 passive solar energy techniques the school could implement.

3. Many scientists are concerned about anthropogenic activities that they believe have contributed to global climate change.

   (A) Name 2 chemicals that can contribute to climate change and describe how these chemicals get into our atmosphere.

   (B) Describe how the chemicals you named above can enhance the greenhouse effect.

   (C) Identify 2 environmental effects of climate change.

   (D) i. In which layers of earth's atmosphere do we find greenhouse gases?

      ii. Describe two ways humans could reduce the release of greenhouse gases.

4. Smog is a concern in many large cities today. Poor air quality affects human health. Acid deposition has been a problem in the past.

(A) Identify and describe the two types of smog.

(B) Name and describe TWO human health or environmental concerns related to breathing smog.

(C) Identify the two predominant chemicals that cause acid deposition and describe the chemical reactions that take place in the atmosphere to cause this problem.

(D) List and explain two problems that acid deposition can cause.

ANSWERS

Chapter 1

Module 1: Environmental Science

Study the Figure: Figure 1.2
Answers will vary

Review Key Terms

1. h
2. g
3. f
4. e
5. b
6. c
7. d

Module 2: Environmental Indicators and Sustainability

Study the Figure: Figure 2.5

1. As carbon dioxide concentrations increase, global temperatures have increased.

2. 1800-1900: $(300 - 275) \div 275 \times 100 = 9\%$

3. 1900-2000: $(400 - 300) \div 300 \times 100 = 33\%$
   The percent change from 1900-2000 is significantly higher than that of previous centuries.

4. Most scientists agree the increase in temperature is due to increased carbon in the atmosphere from burning fossil fuels and deforestation.

Practice the Math: Converting Between Hectares and Acres
50,000 acres = 20,000 hectares
75,000 acres = 30,000 hectares
150,000 acres = 60,000 hectares

Practice the Math: Rates of Forest Clearing
Estimate 1:
15 acre/minute × 0.40 ha/acre = 6 ha/minute
 6 ha/minute × 60 min/1hour = 360 ha/hour
360 ha/hour × 24 hour/day = 8,640 ha/day

Estimate 2:
22,000 acre/day × 0.40 ha/acre = 8,800 ha/day
Estimates 1 and 2 are higher than estimate 3.

Review Key Terms

1. o
2. f.
3. a
4. b

5. g
6. k
7. j
8. m

9. n
10. l
11. e
12. h

13. c
14. d

Module 3: Scientific Method

Study the Figure: Figure 3.1

1. Observation and questioning: Dissolved oxygen levels are higher in shady areas compared to sunny areas. Does temperature affect dissolved oxygen levels in aquatic systems?

2. Testable hypothesis: (Null) Temperature will not affect dissolved oxygen levels within an aquatic system.

3. Temperature readings will be taken in the shady areas and the sunny areas of the pond to determine experimental temperature values. A control tank will be set at room temperature (25 degrees Celsius. The experimental group will include a tank set at 20 degrees Celsius and a second tank set at 30 degrees Celsius. Dissolved oxygen readings will be taken over a period of 5 days for each tank.

Data:

Temperature	Dissolved Oxygen
Control (25 degrees)	8 mg/L
Experimental @20	10 mg/L
Experimental @30	6 mg/L

4. Over the five day period, DO levels were higher in the colder tank (20 degrees) and lowest in the warmer tank (30 degrees). The data indicates temperature does affect oxygen levels in aquatic ecosystems. Higher temperatures result in lower DO levels while cooler temperatures result in higher DO levels. Repeated trials would be necessary to ensure validity of the data.

Review Key Terms

1. b
2. j
3. a

4. d
5. i
6. h

7. k
8. h
9. f

10. g
11. e

Chapter 1 Review Exercises

Check Your Understanding

1. Disciplines incorporated into environmental science include: Biology, ecology, toxicology, atmospheric sciences, chemistry, earth sciences, law, literature and writing, ethics, politics and policy, and economics

2. The 5 key global-scale environmental indicators are: Biodiversity, abundant food production, global surface temperature, the size of the human population and resource depletion

3. Genetic diversity is a measure of the genetic variation among individuals in a population. Species diversity is the number of species in a region or in a particular type of habitat. Ecosystem diversity is a measure of the diversity of ecosystems or habitats that exist in a given region.

4. Anthropogenic activities refer to human activities. Examples include burning fossil fuels, driving cars, cutting down forests, and habitat loss from human construction.

5. The size of the global human population is currently over 7 billion.

6. A person's ecological footprint is a measure of how much a person consumes, expressed in area of land required to support a person's lifestyle.

7. The steps in the scientific method are: Observe and question, form testable hypothesis, collect data, interpret results, disseminate findings.

Practice for Free-Response Questions

Human Activity	Environmental Impact	Environmental Indicator
Increased numbers of human population	Increased use of limited resources/ increased pressures on environmental systems/ increased pollution	Numbers of individuals
Land use changes/ increased urbanization/ agriculture	Increased pollution/ loss of natural habitat/	Water quality/ Deposition rates of atmospheric compounds/ carbon dioxide/ habitat loss rate
Increased rate of species extinctions	Loss of genetic diversity	Biological diversity
Food production	Increased land use/ increased pollution/ increased human population/ habitat loss	Per capita food production/ total food production
Burning fossil fuels	Increased atmospheric carbon dioxide/ increase in global mean temperatures/ sea level rise	Average global surface temperatures/ carbon dioxide concentration/ sea level rise
Overfishing	Declining numbers of fish populations	Fish catch/ fish consumption advisories

Chapter 2

Module 4: Systems and Matter

Study the Figure: Figure 4.7

1. Seawater has a pH of about 8. Lakes affected by acid rain have a pH of about 4. So acidic lakes are about10,000 times more acidic than sea water.

Review Key Terms

1. m	9. k	17. e	25. n
2. o	10. y	18. aa	26. ff
3. d	11. g	19. h	27. w
4. i	12. cc	20. b	28. bb
5. r	13. t	21. f	29. l
6. q	14. a	22. ee	30. x
7. dd	15. z	23. j	31. u
8. c	16. v	24. p	32. s

Module 5: Energy, Flows, and Feedbacks

Practice the Math: Calculating Energy Use and Converting Units (Module 5, p. 46)

1. 1,500 watts + 2,000 watts = 3,500 watts
   3,500 watts ÷ 1000 watts per kw = 3.5 kW
   3.5 kW × 0.5 (30 minutes)= 1.75 kW per 30 minutes
   1.75 kW × $0.10 = 0.175 kW per day
   0.175 kW per day × 7 days per week = $1.23 per week

2. (a) Old model:
   1400 watts × 1 kW/ 1000 watts
   1.4kW × 9 hours/week = 12.6 kWh/week
   12.6 kWh/week × 52 weeks/year = 655.2 kWh/year

   New Model:
   600 watts ×1 kw/1000 watts
   0.6 kW × 9 hours/week = 5.4 kWh/week
   5.4 kWh/week × 52weeks/year = 280.8 kWh/year
   655.2 kWh/year −280.8 kWh/year = 374.4 kWh/year saved by using new model

   (b) 374.4 kWh × $0.10/kWh = $37.44 per year saved
   Cost of new model = $600 − $100 rebate = $500
   $500 ÷ $37.44 = 13.4 years to recover the cost of the dishwasher

Study the Figure: Figure 5.6

1. (35%) × (90%) × (70%) = 22% efficiency

Review Key Terms

1. k	7. i	13. c	19. l
2. e	8. q	14. g	20. o
3. p	9. a	15. t	21. n
4. v	10. b	16. h	22. r
5. m	11. u	17. j	
6. f	12. d	18. s	

Chapter 2 Review Exercises

Check Your Understanding

1. Radioactive decay is the spontaneous release of material from the nucleus. Radioactive decay changes the radioactive element into a different element. For example, uranium-235 decays to form thorium-231.

2. An element's half-life is the time it takes for one-half of the original radioactive parent atoms to decay. It is important to know because some elements that undergo radioactive decay emit harmful radiation. Knowing the half-life will allow scientists to determine the length of time an element may be dangerous.

3. The properties of water are surface tension, capillary action, a high boiling point, and the ability to dissolve many different substances.

4. 100 times… because each step up the scale is a factor of 10 times. So, from 3 to 4 would be 10 times more acidic and from 4 to 5 is another 10 times more acidic. 10 × 10 = 100.

5. An example of potential energy is water stored behind a dam. When the water flows downstream, that potential energy becomes kinetic energy.

6. The first law of thermodynamics is that energy cannot be created or destroyed, so the energy never goes away. The second law of thermodynamics is that the energy is transformed and the amount of energy is the same but it cannot do the same amount of work. Most of the energy is converted into heat.

7. Coal burning power plants are approximately 35% efficient, incandescent light bulbs are about 5 percent efficient and the electrical transmission lines between the power plant and the house is approximately 90 percent efficient.

8. Mono Lake is an example of a negative feedback loop. When the water level drops there is less lake surface area. With less surface area the evaporation rate decreases. Since there is less evaporation the water in the lake slowly returns to its original volume. The system is responding to a change so it will return to its original state. An example of a positive feedback loop is a population growing. More births create a population increase which in turn creates more births. The amplified population is a positive feedback loop.

Practice for Free-Response Questions

Energy from the sun is transformed into chemical energy stored in the tissues of plants through the process of photosynthesis. According the first law of thermodynamics, energy is neither created or destroyed and can be transferred from one type to another. In this case solar energy is transformed in to chemical or potential energy. The second law of thermodynamics explains that the transfer of energy is inefficient. Therefore not all of the solar radiation will be transferred into a usable form within the plant.

Unit 1 Multiple-Choice Review Exam

1. D	8. B	15. B	22. C
2. A	9. B	16. C	23. E
3. C	10. E	17. B	24. D
4. E	11. E	18. E	25. E
5. C	12. A	19. A	26. D
6. D	13. C	20. E	27. A
7. A	14. A	21. D	28. C

Chapter 3

Module 6: The Movement of Energy

Review Key Terms

1. c	7. a	13. e	19. d
2. g	8. q	14. u	20. j
3. n	9. v	15. t	21. o
4. p	10. r	16. i	22. l
5. w	11. h	17. s	23. k
6. f	12. b	18. m	

Module 7: The Movement of Matter

Study the Figure: Figure 7.1

1. Sample answers: **Infiltration:** humans have paved surfaces, which makes them impervious to water and prevents infiltration. **Transpiration:** humans have deforested large tracts of land, which decreases transpiration.

Practice the Math: Raising Mangos

1. (a) 300 trees × $75 = $22,500
   (b) 300 trees × 2 gallons per tree = 600 gallons per day
       600 gallons per day × 365 days per year = 219,000 gallons of water per year
   (c) $1.25 per sapling × 300 saplings = $375

2. (a) Total annual income desired: $200/person × 300 people = $60,000
       Number of trees to produce $60,000 in annual income: $60,000 ÷ $50/tree = 1,200 trees
   (b) 1,200 trees × 10 m^2 = 12,000 m^2
       12,000 m^2 × 0.0001 ha/ 1 m^2 = 1.2 ha
   (c) 1,200 trees × 15 L/ day × 180 days/ year = 3,240,000 L/year

Study the Figure: Figure 7.2

Biological	Geological
Photosynthesis (retains)	Combustion (release)
Cell Respiration (release)	Extraction (release)
Decomposition (release)	Sedimentation (retains)
	Exchange (both)
	Burial (retains)

Study the Figure: Figure 7.3

Process	Product
Nitrogen Fixation	Ammonia, followed by ammonium
Nitrification	Nitrite, Nitrate
Assimilation	Proteins
Denitrification	Nitrate, Nitrous oxide, Nitrogen gas
Mineralization	Ammonium

Review Key Terms

1. p
2. i
3. l
4. r
5. f

6. g
7. t
8. j
9. h
10. s

11. a
12. e
13. k
14. m
15. b

16. q
17. d
18. c
19. o
20. n

Module 8: Responses to Disturbances

Review Key Terms

1. e
2. c
3. a

4. d
5. f
6. b

Chapter 3 Review Exercises

Check Your Understanding

1. Solar energy + 6 H_2O + 6 CO_2 → $C_6H_{12}O_6$ + 6 O_2

2. Energy + 6 H_2O + 6 CO_2 ← $C_6H_{12}O_6$ + 6 O_2

3. Of the total biomass available at a given trophic level, only about 10 percent can be converted into energy at the next higher trophic level.

4. The fast part of the carbon cycle refers to carbon that is associated with living organisms. This part of the cycle flows through the environment. The slow part of the carbon cycle refers to carbon that is held in rocks, in soils, or as petroleum. This carbon may be stored for millions of years.

5. Nitrogen fixation produces ammonia. Nitrification produces nitrates and nitrites. Assimilation produces proteins. Ammonification produces ammonia. Denitrification produces nitrogen gas.

Practice for Free-Response Questions

System	Ecosystem Service
Biological energy transfer: Photosynthesis/ Cell respiration	Carbon sequestration/ carbon cycling
Decomposition	removal of dead matter and return of nutrients to soil
Hydrologic Cycle	Water purification/ moderation of climate
Carbon Cycle	Fossil carbon used for energy/ fuels
Nitrogen Cycle	Bacteria transform nitrogen into usable form to increase crop productivity
Phosphorus Cycle	Bats produce phosphorus rich guano to improve crop productivity

Chapter 4

Module 9: The Unequal Heating of Earth

Review Key Terms

1. c
2. e

3. a
4. b

5. d

Module 10: Air Currents

Study The Figure: Figure 10.6

1. Atmospheric convection currents are global patterns of air movement that are initiated by the uneven heating of the earth. The air moves up into the atmosphere and then returns to the surface of Earth. Hadley cells are convection currents operate between the equator and 30 degrees North and 30 degrees South. Within the Hadley cells air circulates toward the equator near Earth's surface. In the Northern Hemisphere, air moves from north to south. In the Southern Hemisphere, air moves from south to north. Polar cells are convection currents that form at 60 degrees South and 60 degrees South. Air sinks at the poles. The third type of convection current called Ferrell cells lies in between the polar and the Hadley cells. Air currents at this latitude does not form distinct convection currents. None of the convection currents fall in a straight line. Because of the Coriolis effect, the winds bend with the rotation of the earth. From the equator to the Tropic of Cancer winds originate from the northeast and flow in a southeasterly direction. From the equator to the Tropic of Capricorn, the winds originate from the southeast and flow in a northeasterly direction. Because the earth rotates faster at the mid-latitudes, the winds that would otherwise flow directly north or directly south bend to the east.

2. See Figure 10.6.

Review Key Terms

1. g
2. d
3. j

4. e
5. a
6. b

7. k
8. c
9. h

10. f
11. i

Module 11: Ocean Currents

Study the Figure: Figure 11.2

1. Cold water

2. Salt water

3. The thermohaline circulation is driven by the increase in dense salt water in the North Atlantic. This salty water sinks mixing with deep cold water. At the equator water is warmed. The sinking of cold salty water in the North and the rising of the warm water at the equator results in the slow moving deep ocean current.

4. The thermohaline circulation is responsible for moving heat and nutrients around the globe. The thermohaline circulation moderates climates of certain landmasses where warm water warms the climate.

Review Key Terms

1. b   2. a   3. d   4. c

Module 12: Terrestrial Biomes

Study the Figure: Figure 12.4

1. Example 1: The precipitation patterns are most similar to a tropical rainforest. However, because the temperatures are so low, one can conclude that the climatogram most likely represents the boreal forest. Furthermore the growing season (August through October) is similar to the boreal forest. Example 2: Again the temperature pattern suggests this climatogram represents the boreal forest. The growing season appears to be shorter due to limited precipitation rather than temperature. If temperatures were warmer, the climatogram would look more like the temperate rain forest in that the simmer months see increased temperatures and decreased precipitation amounts.

2. Answers will vary.

Review Key Terms

1. i   4. a   7. k   10. c
2. e   5. g   8. d   11. j
3. f   6. b   9. l.   12. h

Module 13: Aquatic Biomes

Review Key Terms

1. a   6. c   11. q   16. n
2. h   7. m   12. o   17. i
3. p   8. g   13. d   18. l
4. f   9. k   14. j
5. r   10. b   15. e

Chapter 4 Review Exercises

Check Your Understanding

1. Unequal heating of Earth by the Sun, atmospheric convection currents, the rotation of Earth, Earth's orbit around the Sun on a tilted axis, and ocean currents.

2. See Figure 9.1.

3. Coriolis effect, atmospheric convection currents, and the mixing of air currents in the mid-latitudes.

4. There are many ways oceans regulate the temperature of Earth. First, gyres, or large-scale water circulation, redistribute heat in the atmosphere. Also, when surface currents diverge, it causes deeper waters to rise and replace the water that has moved away and this water is generally colder. Thermohaline circulation is another way, which is the sinking of dense, salty water in the North Atlantic moving cold water around the world. Finally, every 3 to 7 years, the surface currents in the tropical Pacific Ocean reverse direction allowing warm water from the equator to move, known as El Niño-Southern Oscillation.

5. The following list describes the major biomes on Earth.

   - The tundra is cold and treeless. There is little precipitation. Both temperature and precipitation limit plant growth. Therefore most vegetation is low-growing, such as mosses, lichens and heaths.

   - The boreal forest (taiga) has very cold temperatures and low precipitation. The plant life includes coniferous evergreen trees that can tolerate the cold winters.

   - The temperate rainforest is a coastal biome that experiences moderate temperatures and high precipitation. These biomes support very large coniferous trees including spruce, cedar and redwoods.

   - The temperate seasonal forest has moderate temperatures and moderate amounts of precipitation. This biome supports broadleaf deciduous trees such as beech, maple, oak and hickory.

   - The woodland/ shrubland biome is a temperate biome characterized by hot, dry summers and mild rainy winters. There is a 12-month growing season, but precipitation limits growth. Plants are adapted to fire and drought and include shrubs such as scrub oak, sage and yucca.

   - The temperate grassland has cold/harsh winters and hot/ dry summers. Growth is constrained by cold temperatures in the winter and limited precipitation in the summers. Plant life includes grasses and non-woody flowering plants.

   - The tropical rainforest is a warm and wet biome characterized by little temperature variation and high precipitation. These forests have several distinctive layers of vegetation to include large trees that form a canopy and shorter trees that make up the understory. Epiphytes and woody vines also characterize this biome's plant life.

- The tropical seasonal forests and savannah have warm temperatures and distinct wet and dry seasons. Vegetation ranges form dense stands of trees to relatively open landscapes that include grasses and scattered deciduous trees.

- The subtropical desert is characterized by hot and extremely dry conditions. Vegetation is sparse. Cacti, euphorbs, and succulents are the dominant plant life of this biome

6. The major aquatic biomes on Earth include streams, rivers, lakes, ponds, freshwater wetlands, salt marshes, mangrove swamps, intertidal zone, coral reefs, the open ocean.

Practice for Free-Response Questions

Convection currents are driven by warm and cool air. Warm air rises and as it does it experiences lower atmospheric pressures and adiabatic cooling. The cooling causes condensation and cloud formation which leads to precipitation. As dry air rises it becomes cooler and becomes denser, which causes it to sink. As the air sinks it experiences adiabatic heating. By the time this air reaches the surface of Earth it is hot and dry. Hadley cells cause high precipitation to occur at the equator and hot dry air falls at 30 degrees north and south of the equator leading to deserts at these locations. Ferrell cells and polar cells allow for tropical and polar air to mix in the temperate regions. Because these currents determine temperature and precipitation patterns, they directly influence the location of deserts, grasslands and forests around the globe.

Chapter 5

Module 14: The Biodiversity of Earth

Practice the Math: Measuring Species Diversity

1. Community A
   $H = -[(0.20 \times \ln 0.20) + (0.20 \times \ln 0.20) + (0.20 \times \ln 0.20) \times (0.20 \times \ln 0.20) + (0.20 \times \ln 0.20)]$

   $H = -[(-0.60) + (-0.60) + (-0.60) + (-0.60) + (-0.60)] = 3.0$
2. Community B
   $H = -[(0.05 \times \ln 0.05) + (0.05 \times \ln 0.05) + (0.05 \times \ln 0.05) + (0.05 \times \ln 0.05) + (0.35 \times \ln 0.35) + (0.35 \times \ln 0.35)]$

   $H = -[(-0.15) + (-0.15) + (-0.15) + (-0.15) + (-0.37) + (-0.37)] = 1.34$

Review Key Terms

1. c   3. b
2. a

Module 15: How Evolution Creates Diversity

Study the Figure: Figure 15.3

1. The Afghan hound and the Saluki breeds would be most different from the ancestral wolf. The Chinese shar-pei and Basenji would be most similar. Because the Chinese shar-pei and Basenji are diverge at two branches from the wolf indicating they are most closely related. The Afghan hound and Saluki breeds are most different from the wolf because there are 5 points of divergence (nodes) from the wolf.

Review Key Terms

1. l	6. d	11. f	16. k
2. g	7. j	12. q	17. e
3. n	8. c	13. h	
4. p	9. b	14. m	
5. a	10. o	15. i	

Module 16: Speciation and the Pace of Evolution

Review Key Terms

1. b	3. a	5. e
2. d	4. c	

Module 17: Evolution of Niches and Species Distributions

Study the Figure: Figure 17.1

1. Low pH values (<6) and high pH values (>8) result in smaller populations indicating mortality increases at low and high pH values.

2. Population numbers will decline.

3. Graph:

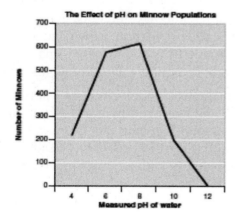

4. The optimal range of tolerance is a pH of 6-8 for the minnows.

Chapter 5 Review Exercises

Check Your Understanding

1. Species evenness measures the relative proportions of species in an area compared to the total number of species in that area. For example, if a community has 25% of species A, 25 percent of species B, 25 percent of species C, and 25 percent of species D, the community is very even. Species richness refers to the number of different species in a given area. High richness means there are a lot of different species. Low richness means there are few different kinds of species in the area.

2. The key ideas the theory of evolution by natural selection are: individuals produce an excess of offspring, not all offspring can survive, individuals differ in their traits, differences in traits can be passed on from parents to offspring and differences in traits are associated with differences in the ability to survive and reproduce.

3. The five random processes of evolution are:

   - **Mutation** describes a random change in the genetic code produced by a mistake in the copying process. If a mutation is not lethal it might increase the organism's chance for survival and reproduction. Organisms with a mutation that improves fitness can pass on the mutation to future generations and the mutation will increase in frequency. Mutations add genetic variation to a population, which increases genetic diversity.

   - **Gene flow** describes the process by which individuals move from one population to another and thereby alter the genetic composition of both populations. New alleles may be introduced into a population by immigrating individuals or some alleles may be removed from a population by emigrating individuals. Either way migrating individuals alter the allele frequency in a population. Gene flow can be helpful in bringing genetic variation to a population that lacks it.

   - **Genetic drift** change in the genetic composition of a population over time as a result of random mating. Genetic drift has a particularly significant effect on the genetic composition of small populations. Random mating among individuals can eliminate some individuals with rare genotypes because they did not find a mate in a given year. Since their genes will not be passed on, the genetic composition of the population will change.

   - **Bottleneck effect** describes a reduction in the genetic diversity of a population caused by a reduction in its size. The reduction of a population can reduce genetic variation and therefore change the genetic composition of a population. Low genetic variation in a population can cause problems such as low fertility and disease. Small populations with low genetic diversity are less able to adapt to changing environmental conditions.

   - **Founder effect** describes a change in the genetic composition of a population resulting from descending from a small number of colonizing individuals. When individuals from the mainland colonize a new island, these founding individuals possess all the alleles for the new population. Because they may not bring all the diversity of the mainland with them, the founders may give rise to a population with a more limited genetic composition.

4. The figure is showing how an original field mouse population has been separated by a river and can no longer breed with one another. This caused two different and genetically distinct populations to form that may not be able to interbreed.

5. The factors that determine the pace of evolution are: if the rate of the environmental change is relatively slow, if the population has high genetic variation for selection to act on, if the population is relatively small, and if the generation time is short.

6. Scientists believe the five causes are habitat destruction, overharvesting, introductions of invasive species, climate change, and emerging diseases.

Practice for Free-Response Questions

A fish population that preys on amphipods is lost due to hypoxia. The larger amphipods are no longer eaten by the fish and can survive lower oxygen levels better than the smaller amphipods. Because environmental conditions favor the larger amphipods they will survive and reproduce more frequently. Over time the larger amphipods will dominate the population (allele frequency shifts).

Unit 2 Multiple-Choice Review Exam

1. E	9. E	17. C	25. D
2. A	10. B	18. E	26. B
3. D	11. B	19. B	27. D
4. A	12. E	20. C	28. E
5. B	13. A	21. B	29. E
6. D	14. B	22. B	30. A
7. C	15. D	23. A	
8. A	16. C	24. B	

Chapter 6

Module 18: The Abundance and Distribution of Populations

Review Key Terms

1. e	4. f	7. a	10. d
2. g	5. c	8. b	11. k
3. l	6. h	9. i	12. j

Module 19 Population Growth Models

Study the Figure: Figure 19.2

1. Initial growth of a small population can be exponential because resources are abundant. When the population reaches about half the carrying capacity growth begins to slow because resources become more limited. At carrying capacity resources cannot support any further growth and the population remains a constant size. The shape of this growth pattern—from a small size, to exponential growth, to slowing growth, and finally to zero growth—is like the letter "S".

Practice the Math: Calculating Exponential Growth

1. Year 1 $N = 15 \times e^{0.25 \times 1}$
   $N = 15 \times 1.28$
   $N = 19$ deer

   Year 5 $N = 15 \times e^{0.25 \times 5}$
   $N = 15 \times 3.49$
   $N = 52$ deer

   Year 10 $N = 15 \times e^{0.25 \times 10}$
   $N = 15 \times 12.18$
   $N = 183$ deer

2. Year 1 $N_t = N_0 e^{rt}$
   $N_t = 100 \times e^{0.75 \times 1}$
   $N_t = 100 \times e^{0.75}$
   $N_t = 100 \times 2.12$
   $N_t = 212$ mosquitoes

   Year 5 $N_t = N_0 e^{rt}$
   $N_t = 100 \times e^{0.75 \times 5}$
   $N_t = 100 \times e^{3.75}$
   $N_t = 100 \times 42.5$
   $N_t = 4252$ mosquitoes

   Year 10 $N_t = N_0 e^{rt}$
   $N_t = 100 \times e^{0.75 \times 10}$
   $N_t = 100 \times e^{7.5}$
   $N_t = 100 \times 1808$
   $N_t = 180,804$ mosquitoes

Review Key Terms

1. m	6. k	11. r	16. g
2. f	7. a	12. e	17. l
3. q	8. o	13. n	18. i
4. j	9. d	14. p	
5. b	10. c	15. h	

Module 20 Community Ecology

Review Key Terms

1. a	5. n	9. f	13. e
2. l	6. j	10. c	14. h
3. d	7. b	11. i	
4. g	8. m	12. k	

Module 21 Community Succession

Study the Figure: Figure 21.4

1. Species diversity increases with size; larger islands have greater diversity. In addition, species diversity decreases with distance from a mainland. Therefore, the larger the island and the closer it is to the mainland, the greater the species diversity.

Review Key Terms

1. e	3. b	5. a
2. c	4. d	

Chapter 6 Review Exercises

Check Your Understanding

1.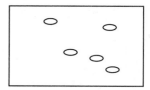

Random   Uniform   Clumped

2. The availability of food is an example of a density-dependent factor. Hurricanes, tornadoes, floods, fires, and volcanic eruptions are examples of density-independent factors.

3. J shaped: See Figure 19.1; Logistic: See Figure 19.2.

4. See Figure 19.4.

5. As the hares increase in number there is more food for the lynx. As the lynx starts to eat more hares the lynx increase in numbers and the hares decrease in number. This cycle continues.

6.

Relationship	Definition	Example
Mutualism	In this species interaction, both species benefit.	Acacia trees and *Pseudomyrmex* ants
Commensalism	In this species interaction, one species benefits, but the other is neither helped nor harmed.	Fish that use coral reefs for hiding places
Parasitism	In this species interaction, one species benefits while the other is harmed.	Tape worms that live inside the intestine of animals

7. Primary succession begins with bare rock, devoid of soil. This type of succession requires a long period of time since soil formation must occur. Organisms such as algae, lichens and mosses colonize these areas of primary succession first. Secondary succession occurs after an area has experienced a disturbance such as a forest fire. Soil remains and therefore this community experiences rapid colonization of plant species such as grasses and wildflowers.

Practice for Free-Response Questions

1. See Figure 19.2. *K*-selected species experience low intrinsic growth rates that cause a population to increase slowly until it reaches carrying capacity. Populations of *K*-selected strategists are primarily regulated by density-dependent factors.These patterns are reflected in the logistic model.

2. See Figure 19.3. *r*-selected strategists experience high intrinsic growth rates as reflected in the exponential growth model. These populations experience population over-shoots and die-off events. Populations of *r*-selected species are regulated by density-independent factors.

Chapter 7

Module 22 Human Population Numbers

Practice the Math: Calculating Population Growth

$$\frac{(15\text{-}5) + 84}{1000} = \frac{x}{5,900,00}$$

$$\frac{94}{1000} = \frac{x}{5,900,00}$$

$$1000x = 94 \times 5,900.000$$

$$1,000x = 94 \times 5,900,000$$

$$x = 554,600 \text{ people per year}$$

$$\text{Growth Rate} = \frac{554,600}{5,900,000} \times 100 = 9.4\%$$

$$\text{Doubling Time} = \frac{70}{9.4} = 7.4 \text{ years}$$

Study the Figure: Figure 22.8

1.

- India has high numbers of youth and lower numbers of older people. This country is positioned for rapid population growth because it has a large population of individuals in pre-reproductive years.

- The United States is experiencing stable population size because it has equal numbers of individuals in younger and older age groups.

- Germany has a greater number of older people than younger people. Therefore, Germany's population would be expected to decline.

- China's population is also expected to decline since there are fewer young people than older people.

Review Key Terms

1. o	6. k	11. g	16. c
2. f	7. a	12. n	17. e
3. m	8. i	13. r	18. l
4. h	9. p	14. d	
5. j	10. b	15. q	

Module 23 Economic Development, Consumption, and Sustainability

Study the Figure: Figure 23.4

1. Total fertility rate is the number of children a woman will have in her lifetime. As educational opportunities increase for women, TFR decreases. With increased education, opportunities for work increase which may lead to women having fewer children and delaying childbearing.

Review Key Terms

1. b   3. c   5. a
2. d   4. f   6. e

Chapter 7 Review Exercises

Check Your Understanding

1. $(15 + 5) - (10 + 2) \div 10 = (20 - 12) \div 10 = 8 \div 10 = 0.8\,\%$

2. Doubling time $= 70 \div 0.8 = 87.5$ years

3. Rapid growth, see Figure 22.8(a); Stable growth, see Figure 22.8(b); Declining growth, see Figure 22.8(c)

4. See Figure 23.

5. IPAT: Impact = population × affluence × technology. According to the IPAT equation, the three major factors that influence environmental impact are the size of the population, the population's wealth, and its technological advancement. The larger the IPAT, the larger the environmental impact.

6. The four types of economic activity considered in GDP are consumer spending, investments, government spending, and exports minus imports.

Practice for Free-Response Questions

Diagram	Population Characteristics
Population pyramid	• This population is probably in the second stage of the demographic transition because of the large numbers of young people indicated at the base of the pyramid. • While improved sanitation and health care may be available, education for women is limited. • The population has a high birth rate and lowered death rate which positions the country for exponential growth.
Even age distribution	• Even age distributions indicate the population is experiencing slow or stable growth. • This population is probably in the third phase of the demographic transition model. • The economy and education levels within this population have improved. • Increased affluence, education and greater access to birth control decreases birth rates to more evenly match death rates.
Inverted Pyramid	• This population is experiencing declining growth rates. • Crude birth rates are below death rates and therefore this population has higher numbers of elderly individuals. • High levels of affluence and economic development encourage women to have few to no children. • Fewer people in the work force means the country may face a shortage of health care workers and experience an increased tax burden on people in the work force.

Unit 3 Multiple-Choice Review Exam

1. D	8. A	15. E	22. C
2. B	9. E	16. B	23. B
3. C	10. D	17. D	24. D
4. A	11. B	18. A	25. B
5. B	12. A	19. E	26. D
6. E	13. C	20. B	27. C
7. C	14. A	21. D	28. C

Chapter 8

Module 24: Mineral Resources and Geology

Practice the Math: Plate Movement

1. 500 km = 500,000 m = 500,000,000 mm
   500,000,000 mm ÷ 15mm = 33,333,333 years
   It will take about 33 million years for Shaky Acres to reach New Hanover.

2. 718 km = 718,000 m = 718,000,000 mm
   718,000,000 mm ÷ 36 mm/year = 19,944,444 years
   It will take approximately 20 million years for Palm Springs to reach San Jose.

Review Key Terms

1. i	8. h	15. bb	22. y
2. w	9. u	16. o	23. g
3. l	10. r	17. x	24. c
4. p	11. a	18. b	25. s
5. j	12. aa	19. q	26. v
6. z	13. n	20. d	27. k
7. e	14. t	21. f	28. m

Module 25: Weathering and Soil Science

Study the Figure: Figure 25.7

1. The O horizon is composed of organic matter.

2. The A horizon is also known as top soil and is a mixture of organic material and underlying mineral material.

3. The E horizon is a zone of leaching where metals and nutrients are moved through and accumulate in the B horizon.

4. The B horizon is a zone of accumulation of metals and nutrients.

5. The C horizon is similar to the parent material and is the least weathered portion of the soil profile.

Study the Figure: Figure 25.8(a)

1. If the composition of a soil is found to be 85% sand, 15 % silt, and 55% clay it is known as sandy clay.

Review Key Terms

1. f	7. x	13. d	19. w
2. b	8. s	14. q	20. h
3. v	9. j	15. t	21. p
4. o	10. a	16. c	22. e
5. r	11. g	17. u	23. n
6. m	12. i	18. k	24. l

Chapter 8 Review Exercises

Check Your Understanding

1. See Figure 24.3(a)

2. the theory of Plate tectonics states that Earth's lithosphere is divided into plates, most of which are in constant motion. New lithosphere is added at spreading zones and older lithosphere is recycled into the mantle at Subduction zones.

3. See Figure 24.8.

4. Each time you move up the Richter scale you are increasing the damage done by an earthquake by a multiple of 10.

5. Weathering and erosion breakdown rock exposed at the earth's surface. Wind, water, chemical and biological agents play an active role in both of these processes. Transport describes the movement of particles and sediment. Finally compression describes the pressure that builds over layers of materials. After long periods of time, this pressure forms sedimentary rock.

6. See Figure 25.7.

7. The five properties that determine soil are parent material, climate, topography, organisms, and time.

8. Physical properties: Texture describes the percentage of sand, silt and clay contained in the soil. Permeability is dependent upon soil texture and describes how quickly soil drains. Chemical properties: Cation exchange capacity (CEC) describes the soil's ability to absorb and release cations. It is also referred to as the soil's nutrient holding capacity. Base saturation is the proportion of soil bases to soil acids, expressed as a percentage. Generally soil bases encourage plant growth while soil acids are detrimental to plant nutrition. Biological properties: Soil organisms populate the soil. Microorganisms including bacteria, fungi, and protozoans act as nutrient cyclers and decomposers. Larger organisms including earthworms and rodents contribute to soil mixing and breaking down large organic material.

9. Figure 25.8 shows how sand is very porous and therefore water flows through it quickly. Silt is less porous so water flows through it slower than sand. Clay is impermeable and therefore holds water and the water does not move through it.

10. Surface mining refers to mining techniques used to remove minerals or ores that are deposited close to the surface. Subsurface mining refers to mining techniques used to remove a resource more than 100 m below Earth's surface.

Practice for Free-Response Questions

Mining Process	Process Defined	Environmental Impacts
Surface Mining	• A variety of techniques used at the surface to extract valuable metals and minerals. • Strip mining: removal of strips of soil and rock to expose ore. • Open-pit mining: a large visible pit in the ground to extract resources that are close to the surface but also extend vertically and horizontally. • Mountain top removal: the entire top of the mountain is removed with explosives to expose mineral resources. • Placer mining: process of looking for mineral, metals and precious stones in river sediments.	• Increased atmospheric dust. • Contamination of water that percolates through tailings. • Habitat destruction and fragmentation of land.
Subsurface Mining	• Mining techniques used when the desired resource is more than 100 m below the surface of Earth. Tunnels are dug into mountains and miners are delivered to the worksite by elevator.	• Emissions from fossil fuels used in mining equipment. • Acid mine drainage contaminate soil and nearby waterways. • Water that percolates through tailings is also contaminated. • Road construction to mines fragments habitat.

Chapter 9

Module 26: The Availability of Water

Review Key Terms

1. j	4. a	7. h	10. g
2. f	5. e	8. c	11. i.
3. b	6. d	9. k	

Module 27: Human Alteration of Water Availability

Review Key Terms

1. e
2. g
3. i

4. c
5. a
6. b

7. d
8. f
9. h

Study the Figure: Figure 27.5

1. The Aral Sea has declined in surface area by 60 percent because of water diversion. Water diversion decreased fresh water inputs and therefore increased salinity of the remaining lake water destroying fish populations. Dust storms have eroded soil. Salt, pesticide residues and dust have polluted the atmosphere. Local climate has been because water no longer moderates the climate. Summers are hotter and winters are colder.

Study the Figure: Figure 27.6

1. Distillation heats saltwater to a boiling point and then runs cool seawater through a coil to cause steam to condense. The salt-free condensation is collected and flows out of the chamber. One drawback for this process is that it leaves a great deal of salt behind which can be environmentally costly. In addition, the process uses a great deal of energy.

2. Reverse osmosis uses pressure to force seawater through a semi-permeable membrane. Pure water passes through but salt does not. Salt-free water is collected and flows from the chamber. While this process is less energy intensive, it does produce a great deal of brine which can be harmful to plant and animal life.

Module 28: Human Use of Water Now and in the Future

Practice the Math: Selecting the Best Washing Machine

1. a. 8 loads × 4 weeks = 32 loads per month

    b. (200 L – 100 L/load) × 32 loads per month = 3,200 L saved per month
       6,400 L – 3,200 L = 3,200 L of water saved

    c. $\frac{\$0.35}{1,000\text{ L}}$ × 3,200 L/month = $1.12 savings per month

2. a. 150 uses/ year × 35 L/ use = 5,250 L/ year

    b. 5,250 L/ year × $0.50/ 1000L = $2.63

    c. ($500 – $250) ÷ $2.63/year = 95 years

Review Key Terms

1. c   2. b   3. a   4. d

Chapter 9 Review Exercises

Check Your Understanding

1. An unconfined aquifer is an aquifer where water can easily flow in and out. A confined aquifer has an impermeable, or confining, layer which impedes water flow to or from the aquifer.

2. A saltwater intrusion is an infiltration of salt water in an area where groundwater pressure has been reduced from extensive drilling of wells. The saltwater moves into the aquifer and contaminates well water. It is common in coastal areas. An intrusion occurs when too many wells are drilled along a coastline; the water table lowers rapidly and the pressure in the aquifer is reduced, and saltwater from the nearby ocean moves into the aquifer. Managing the number of wells drilled can prevent the reduction in pressure, and the saltwater will be less likely to intrude.

3. During heavy rainfall events that might otherwise lead to flooding, wetlands absorb and release water slowly, reducing the likelihood of flooding and saving property and therefore providing economic benefits to humans.

4. The construction of impermeable and paved surfaces (buildings and pavement) contributes to flooding. The surfaces do not allow water to infiltrate into the ground. This causes the water to run off into storm sewers or streams.

5. Benefits of dams include water for human consumption, generation of electricity, flood control, and recreation. Consequences of dams include having to displace people, disrupting a habitat by flooding to create a reservoir, interruption of the water flow that many organisms need, and loss of breeding grounds for animals such as salmon.

6. 70 percent of water is used for irrigation of crops, 20 percent for industry and 10 percent for household uses.

Practice for Free-Response Questions

1. Humans have built dikes and levees to prevent flooding on adjacent land. Fertility of adjacent land is reduced due to decreased deposition of sediments. Flooding may be prevented in one area but may occur in another.

2. Humans have built dams to restrict water flow in a river or stream. Migration of fish may be altered and populations may be impacted. Fish populations may also be impacted when water is released through turbines. Loss of seasonal flooding limits natural disturbances that encourage colonization of certain species. Habitats may be fragmented. Decreased flow of water downstream and flooding upstream drastically changes ecosystems.

3. Aqueducts are constructed to carry water from one location to another. Construction of aqueducts can disrupt ecosystems and fragment an environment. Since water is diverted away surface waters, those ecosystems may be disrupted if overdrawn.

4. Desalinization of salt water provides fresh drinking water to water-poor areas. Environmental consequences include waste salt and brine that can pollute soil and change salinity areas in sensitive marine ecosystems. The desalinization process is energy intensive. Therefore if fossil fuels are used to drive the process, emissions may contribute to air pollution.

Unit 4 Multiple-Choice Review Exam

1. B	8. C	15. B	22. C
2. A	9. E	16. E	23. E
3. C	10. A	17. C	24. D
4. B	11. D	18. A	25. A
5. D	12. D	19. D	26. E
6. E	13. C	20. B	27. C
7. E	14. E	21. A	

Chapter 10

Module 29: Land Use Concepts and Classification

Review Key Terms

1. a	3. c	5. d
2. e	4. b	

Study the Figure: Figure 29.3

1. When a population is small, growth is relatively slow. As a population reaches carrying capacity, growth also slows. Half-way in between these two points the graph indicates exponential growth of a population, which is very fast. Therefore, harvesting to this point will allow populations to recover more quickly while still providing useful amounts of a natural resource.

Module 30: Land Management Practices

Review Key Terms

1. t	8. x	15. o	22. bb
2. j	9. f	16. q	23. d
3. p	10. b	17. s	24. e
4. m	11. z	18. u	25. g
5. c	12. a	19. v	26. i
6. h	13. y	20. w	27. k
7. n	14. r	21. aa	28. l

Study the Figure: Figure 30.2

1. Clear cutting involves removing almost all of the trees within a given area. Because the area is often replanted with the fast growing tree species and trees will be exposed to large amounts of light, clear cutting can result in large groups of trees reaching their maximum sustainable yield in a short amount of time. Disadvantages to this practice are numerous and include habitat loss and fragmentation, loss of biodiversity, increases in wind and water erosion, loss of soil and nutrients, increased turbidity in nearby aquatic systems and lowed aesthetic value of the forest.

2. Selective cutting removes single trees or relatively small numbers of trees among many in a forest. Benefits to this practice include increasing sunlight for smaller trees allowing different aged stand of trees to thrive increasing growth rates of tree populations. This practice is ecologically less destructive when compared to clear cutting. However, there are still environmental impacts associated with selective cutting to include construction of logging roads which fragment forests, impact species diversity, cause soil compaction and lead to a loss of nutrient absorption and water infiltration.

Chapter 10 Review Exercises

Check Your Understanding

1. Change to the landscape is the single largest cause of species extinctions.

2. Answers will vary. If the baker next door begins work 3am every morning, the noise may interrupt your sleep and you will not be productive the following day. The interruption of your sleep is a negative externality caused by the bakery running its business.

3. Some of the uses of land include timber production, defense, urban, residential, transportation, recreational and wildlife lands, cropland, forest grazing land, and grassland/grazing land.

4. When fires are suppressed, the accumulation of large quantities of dead biomass can accumulate. Eventually when a fire does begin, it can become difficult to control and lead to property damage and loss of life.

5. There has been increased movement of people from the city to surrounding rural areas. Increased availability of cars has led to increased construction of highways. Increased emissions from fossil fuels and habitat fragmentation are consequences of urbanization. Because people leave the city, revenue from property, taxes and sales begin to shrink. Faced with decreased city revenue, cities are forces to reduce services. This may lead to increased crime, deteriorating infrastructure and higher taxes. As the population shifts from the city to the suburbs, jobs follow. This cascade effect leading to the degradation of city social environments is referred to as urban blight.

Practice for Free-Response Questions

Maximum sustainable yield is the maximum amount of a resource that can be harvested without compromising the future availability of the resource. This amount is most often at half the carrying capacity. At this point the population is growing exponentially. At this point the resource can be replenished quickly due to rapid population growth. Allowing the maximum amount of a population to be harvested rather than entirely prohibiting the practice allows for human needs to be met and economic growth to occur.

Chapter 11

Module 31: Human Nutritional Needs

Review Key Terms

1. e	3. g	5. h	7. c
2. b	4. f	6. a	8. d

Module 32: Modern Large-Scale Farming

Practice the Math: Land Needed for Food

1. 2,200 kilocalories/day × 30 days/month = 66,000 kilocalories/month

2. 2,200 kilocalories/day ÷ 53 kilocalories/apple = 41.5 apples/day
   66,000 kilocalories/month ÷ 53 kilocalories/apple = 1,245 apples/month
   or 41.5 apples/day × 30 days = 1,245 apples/month

3. 6,800,000,000 people × 41.5 apples/day = 2.8×10^{11} apples/day

4. 2,000 kilocalories/ day × 365 days/ year = 730,000 kilocalories/ year
   40 bushels/ ha × 27 kg/ bushel × 3500 kcal/kg = 3,780,000 kcal/ ha
   730,000 kcal/year ÷ 3,780,000 kcal/ha = 0.19 ha

Study the Figure: Figure 32.6

1. The pesticide treadmill is a positive feedback loop. Because pests become resistant to pesticides over time, farmers must use greater amounts and stronger pesticides must be developed to maintain effective control.

Review Key Terms

1. u	7. r	13. v	19. f
2. i	8. c	14. s	20. h
3. m	9. l	15. q	21. d
4. e	10. a	16. o	22. b
5. g	11. t	17. n	
6. p	12. j	18. k	

Module 33: Alternatives to Industrial Farming Methods

Review Key Terms

1. h	5. e	9. o	13. a
2. j	6. m	10. g	14. d
3. n	7. k	11. c	15. b
4. l	8. i	12. f	

Chapter 11 Review Exercises

Check Your Understanding

1. Reasons for undernutrition and malnutrition include poverty, political and economic factors, political unrest, rise in food prices, and food being diverted to feed livestock and poultry.

2. Most energy subsidies in modern agriculture go to producing fertilizers and pesticides, operating tractors, pumping water for irrigation, harvesting, and transportation."

3. Irrigation can cause waterlogging, which occurs when soil remains under water for prolonged periods. Waterlogging prevents roots from receiving enough oxygen and impairs plant growth. Irrigation can also lead to salinization, which occurs when the small amount of salts in irrigation water become concentrated on the surface of the soil through evaporation. or when small amounts of salts in irrigation water become highly concentrated on the soil surface through evaporation (salinization).

4. Disadvantages of synthetic fertilizers include the use of fossil fuels, runoff of the fertilizer into bodies of water which can reduce oxygen levels, and they do not add organic matter to the soil.

5. Shifting agriculture involves clearing the land and using it for only a few years until the soil is depleted of nutrients. The traditional method uses a technique called "slash-and-burn," in which existing trees and vegetation are cut down, placed in piles, and burned. The resulting ash is rich in nutrients, which makes the soil more fertile. However, these are usually depleted quickly and in areas of heavy rainfall, the nutrients may be washed away, along with some of the soil, which further reduces nutrient availability. After a few years the farmer usually moves on to another plot. Once the soil has been removed or depleted of nutrients it is difficult for anything to grown back with harms the biodiversity leading to habitat loss in the area.

6. Sustainable agriculture fulfills the need for food and fiber while enhancing the quality of the soil, minimizing the use of non-renewable resources, and allowing economic viability for the farmer.

7. There is evidence that antibiotics given to confined animals are contributing to an increase in antibiotic-resistant strains of microorganisms that can affect humans. Another concern is waste disposal of the manure. Runoff into waterways can also be a problem.

Practice for Free-Response Questions

DDT is a persistent pesticide that remains in the environment for a long time. DDT was banned in part because it was found to accumulate in the fatty tissues of animals, such as eagles and pelicans. (See Chapter 17.) The concentrations of DDT in these animals caused them to lay eggs with thin shells that easily cracked during incubation.

Unit 5 Multiple-Choice Review Exam

1. A	8. C	15. A	22. B
2. D	9. D	16. D	23. E
3. D	10. D	17. B	24. C
4. C	11. C	18. E	25. A
5. B	12. E	19. C	
6. E	13. D	20. E	
7. B	14. A	21. D	

Chapter 12

Module 34: Patterns of Energy Use

Study the Figure: Figure 34.1
1. 87 percent of the worldwide annual energy consumption is nonrenewable energy: 82 percent for fossil fuels and 5 percent for nuclear fuel.

Practice the Math: Efficiency of Travel

1. 3.6 MJ/passenger-kilometer ÷ 3 people = 1.2 MJ/passenger-kilometer
   1.2 MJ/passenger-kilometer × 3910 km/trip = 4692 MJ/passenger-trip
   Carpooling is better than traveling by air, alone in a car, and by bus but not better than traveling by train.

Study the Figure: Figure 34.7

1. Pulverizer: breaks up large solid pieces of coal into a dust that can be easily burned

2. Boiler: device in which water is heated to make steam

3. Steam: contains kinetic energy that is transferred to the blades of the turbine

4. Turbine: device with a center shaft that turns the generator

5. Condenser: fixture in which steam is changed back to liquid form (water)

6. Generator: device connected to turbine that creates electrical current

7. Transmission line: wires that carry the energy from the power plant throughout the electric grid

Practice the Math: Calculating Energy Supply

1. $\dfrac{324,000,000 \text{ kWh /month}}{600 \text{ kWh /month /home}} = 540,000 \text{ homes}$

Review Key Terms

1. l	4. h	7. j	10. i
2. f	5. d	8. k	11. g
3. c	6. b	9. a	12. e

Module 35: Fossil Fuel Resources

Study the Figure: Figure 35.7

1. Using the value of 1.5 from the per capita energy use for 2010: $[(1.5-1) \div 1] \times 100\% = 50\%$

2. As per capita energy use has increased and then leveled off, energy use per dollar of GDP continues to slowly decline.

Review Key Terms

1. i	3. d	5. h	7. c	9. f
2. g	4. b	6. a	8. e	

Module 36: Nuclear Energy Resources

Practice the Math: Calculating Half-Lives

1. ½ = 700,000,000 years
   ¼ = 1,400,000,000 years
   ⅛ = 2,100,000,000 years

2. First find the number of half-lives in 1,295 years.
   $$\frac{1{,}295 \text{ years}}{185 \text{ years}} = 7 \text{ half lives}$$

   Half of the remaining material decays after every half-life; therefore, start with the original amount and multiply by one-half for every half-life.

150 g × ½ = 75 g	1 half-life
75 g × ½ = 37.5 g	2 half-lives
37.5 g × ½ = 18.75 g	3 half-lives
18.75 g × ½ = 9.375 g	4 half-lives
9.375 g × ½ = 4.6875 g	5 half-lives
4.6875 g × ½ = 2.34375 g	6 half-lives
2.34375 g × ½ = 1.171875	7 half-lives

Review Key Terms

1. c
2. a
3. e
4. g
5. f
6. b
7. d

Chapter 12 Review Exercises

Check Your Understanding

1. Commercial energy sources, such as coal, oil, and natural gas, are bought and sold. Subsistence energy sources, for example, straw and dung, are gathered and used by individuals for their own needs.

2. Energy consumption has gone up, particularly the use of coal, oil, nuclear, and natural gas. Wood and hydroelectricity have stayed relatively low.

3. More passengers travel on a bus than in a car with a single driver. Therefore the energy expended per person is less on a bus than in a car with a single driver.

4. Coal is burned to transfer energy to water, which becomes steam. This steam is transferred to the blades of a turbine. The shaft in the center of the turbine turns the generator, which generates electricity.

5. Steam used for industrial purposes or to heat buildings can be diverted to turn a turbine first, and create electricity.

6. Vegetation dies and is buried under anaerobic conditions, forming peat. Peat is compressed to form lignite. Further compression yields sub-bituminous and bituminous coal and more pressure and in time forms anthracite.

7. Advantages: Coal is energy-dense, plentiful, easy to exploit by surface mining, the technological demands of surface mining are relatively small, and economic costs are low. Disadvantages: Mining coal is dangerous, coal releases sulfur in the atmosphere, contributes to air pollution, and leaves behind ash that causes environmental problems for local residents.

8. Advantages: Petroleum is convenient to transport and use, it is relatively energy-dense and burns cleaner than coal. Disadvantages: Petroleum releases sulfur, mercury, lead, and arsenic to the atmosphere when burned, oil spills cause major environmental damage.

9. Advantages: Natural gas has fewer impurities and therefore emits almost no sulfur dioxide or particulates when burned, and natural gas emits only 60 percent as much carbon dioxide as coal. Disadvantages: Methane can escape into the atmosphere, the process of drilling and opening the rock can release the gas, and water used in the process can contaminate groundwater.

10. Advantages: Nuclear energy does not produce air pollution, and countries with limited fossil fuel resources can obtain energy independence. Disadvantages: The possibility of accidents and the difficulty of disposing of radioactive waste.

Practice for Free-Response Questions

Possible answers include:

- Store it safely somewhere on Earth indefinitely where it cannot leach into the groundwater or otherwise escape into the environment.
- Store it far away from human habitation in case of any accidents and secure against terrorist attack. Some examples are storage in above ground storage buildings, using alternative energy and reducing the amount of nuclear fuel needed. '

Chapter 13

Module 37: Conservation, Efficiency, and Renewable Energy

Practice the Math: Energy Star

1. $0.10/hour × 10 hours per day = $1.00 savings per day
$1.00 × 200 days = $200
It would take 200 days to recover the extra cost of the Energy Star appliance.

2. (a) Non Energy Star refrigerator. This refrigerator uses 0.5 kWh.
0.5kW × 10 hours/day × 365 days/year × $0.10 = $1,825/year

(b) Energy Star refrigerator. This refrigerator uses 0.4 kWh.
0.4 kW × 10 hours/day × 365 days/year × $0.10 = $1460/year

$1460/year × 5 years = $7300 for five years
$1825/year × 5 years = $9125 for five years
$9125 – $7300 = $1825 in savings over five years

If you purchase the Energy Star model, you will save $1825 over five years.

Study the Figure: Figure 37.7

1. 0.15 (wind energy) × 0.07 (all renewable energy) ×100% = 1.05 %

Review Key Terms

1. b	3. a	5. e	7. f
2. d	4. h	6. h	8. g

Module 38: Biomass and Water

Study the Figure: Figure 38.1

1. Energy sources that do not come from the Sun include tidal, geothermal, and nuclear.

Review Key Terms

1. c	5. d	9. e	13. m
2. g	6. h	10. l	
3. j	7. b	11. f	
4. a	8. k	12. i	

Module 39: Solar, Wind, Geothermal, and Hydrogen

Study the Figure: Figure 39.7

1. China has a large land mass and a high need of energy because of a large population. Therefore, it has more resources, space, and people to harvest wind energy. At the same time because there are more people and more land, the need for energy is much higher. Even though China produces a lot of energy from wind, it must use other types of energy to meet its needs.

2. 25 GW (5 is 20% of 25)

Review Key Terms

1.	e	3.	c	5.	g	7.	h
2.	b	4.	a	6.	d	8.	f

Chapter 13 Review Exercises

Check Your Understanding

1. Answers will vary. Some options for conserving energy include: Lower thermostat during cold months, walk or ride a bike when possible instead of driving, and turn off your computer or other electronic devices when not in use.

2. Some ways to utilize passive solar design in the northern hemisphere would be to construct a house with south-facing windows, use double-paned windows, place windows so they get the most light into the building, use dark materials on the roof to absorb solar energy, build roof overhangs, and use window shades.

3. Modern carbon is carbon currently in biomass and recently in the atmosphere. Fossil carbon is carbon in fossil fuels.

4. Answers will vary. Examples include corn, corn by-products, wood chips, sugar cane, crop waste, switch grass, algae, soybeans, and palms.

5.

Energy Resource	Advantage	Disadvantage
Liquid Biofuels	Potentially renewable, can reduce dependence on fossil fuels	Loss of agricultural land, higher food costs, deforestation
Solid Biomass	Potentially renewable, available to everyone, reduces waste	Indoor/outdoor air pollution, erosion, deforestation, possible net increase in carbon emissions
Photovoltaic solar cells	Nondepletable resource	Manufacturing requires high input of metals and water, geographically limited
Solar water heater systems	Nondepletable resource	Geographically limited, high initial costs
Hydroelectricity	Nondepletable resource, flood control, recreation	High construction costs, threats to river ecosystems, siltation
Tidal energy	Nondepletable resource	Geographically limited, potential disruption to marine organisms
Geothermal energy	Nondepletable resource	Geographically limited, emits hazardous gases and steam
Wind energy	Nondepletable resource, low up front cost	Turbine noise, death of birds and bats
Hydrogen fuel cell	Efficient, zero pollution	Producing hydrogen is energy intensive, lack of distribution network, hydrogen storage challenges

Practice for Free-Response Questions

Dams can prevent fish from being able to swim upstream to spawn. Fish ladders can work but disadvantages are that some fish fail to utilize them and some predators learn to monitor the fish ladders for their prey.

Unit 6 Multiple-Choice Review Exam

1. E	9. A	17. C	25. D
2. B	10. E	18. E	26. A
3. B	11. D	19. A	27. A
4. C	12. B	20. B	28. D
5. C	13. C	21. B	29. B
6. A	14. E	22. E	
7. A	15. D	23. C	
8. D	16. A	24. D	

Chapter 14

Module 41: Wastewater from Humans and Livestock

Study the Figure: Figure 41.6

1. The thickened sludge must be taken to the landfill, burned, or used for fertilizer.

Practice the Math: Building a Manure Lagoon

1. (a) 60 L of manure a day × 460 animals = 27,600 L a day.
   (b) 60 L of manure a day × 460 animals × 7 days/week = 193,200 L a week.
   (c) 60 L of manure a day × 460 animals × 365 days/year = 10,074,000 L a year.

2. Daily manure production = 50 L/animal × 1500 animals= 75,000 L
   Minimum lagoon capacity = 75,000 L × 45 days = 3,375,000 L

3. Number of trips = 3,375,000 L × (1 trip ÷ 40,000L) = 84.375 trips
   Round to 85 trips.

Review Key Terms

1. j	5. d	9. e	13. m
2. n	6. f	10. g	14. o
3. l	7. a	11. h	15. k
4. b	8. c	12. i	16. p

Module 43: Oil Pollution

Study the Figure: Figure 43.2

1. Environmental laws are often stricter and are enforced more heavily, with greater consequences such as fines and requirements to cleanup spills in North America than in many other places.

Review Key Terms

1. b   3. c
2. a

Module 44: Nonchemical Water Pollution

Review Key Terms

1. b
2. a

Module 45: Water Pollution Laws
Review Key Terms

1. c
2. a
3. b

Chapter 14 Review Exercises

Check Your Understanding

1. Point source pollutants can come from a distinct location such as a factory or sewage treatment plant. Nonpoint source pollutants come from a large area such as an entire farming region, a suburban community with many lawns and septic systems, or storm runoff from parking lots.

2. Wastewater dumped into bodies of water naturally undergoes decomposition by bacteria, which creates a large demand for oxygen in the water. Second, the nutrients that are released from wastewater decomposition can make the water more fertile, and third, wastewater can carry a wide variety of disease-causing agents.

3. Underground pipes carry the waste to treatment plant. Large debris is filtered out by screens and sent to landfill. Solid waste sludge settles to the bottom of the tanks. Bacteria break down organic material to CO_2 and inorganic nutrients. Settled particles are added to sludge. Sludge is thickened by removing water and the thickened sludge is taken to the landfill, burned, or used for fertilizer. The water continues and is exposed to chemical or ultraviolet light to kill pathogens. The treated water is then released to a river or lake.

4. Lead can come from pipes of older homes, brass fittings containing lead, and solder. Arsenic is found naturally in the Earth's crust but can get into groundwater from mining and industrial uses such as wood preservatives. Mercury comes from the burning of fossil fuels, incineration of garbage, hazardous waste, medical supplies, and dental supplies.

5. DDT was designed to target insects, but moved up the food chain all the way to eagles that consume fish. Eagles produced eggs with thinner shells that would break during incubation.

6. Thermal pollution can induce thermal shock and kill organisms that are not adapted to higher temperatures. It can also cause organisms to increase their respiration rates. Because. warmer water does not contain as much dissolved oxygen as cold water, an increase in respiration rates in warmer water can also be fatal.

Practice for Free-Response Questions

1. Containment: plastic barriers that float on the water and keep the oil from spreading further.

2. Chemical: applying chemicals that help break up and disperse the oil before it hits the shoreline and causes damage to the coastal ecosystems.

3. Bacteria: genetically engineered bacteria that occur naturally and obtain energy by consuming oil emerging from natural seeps.

Chapter 15

Module 46: Major Air Pollutants and Their Sources

Study the Figure: Figure 46.5

1. Of all the air pollutants produced by on-road vehicles, carbon monoxide is produced in the highest quantity (approximately 45 percent).

2. We know that 50 percent of our electricity generation is fueled by coal and burning coal produces nitrogen oxides. Nitrogen oxides are produced in virtually all combustion processes including burning coal in a power plant.

Review Key Terms

1. d	4. k	7. c	10. g
2. e	5. a	8. j	11. f
3. h	6. i	9. b	

Module 47: Photochemical Smog and Acid Rain

Study the Figure: Figure 47.1

1. (a) $NO_2 \rightarrow NO + O \rightarrow O + O_2 \rightarrow O_3$ (ozone)

    (b) O_3 (ozone) $+ NO \rightarrow O_2 + NO_2$

    (c) $NO_2 \rightarrow NO + O \rightarrow O + O_2 \rightarrow O_3$ (ozone) $\rightarrow$ Photochemical smog or
        $NO_2 \rightarrow NO + O \rightarrow NO + VOCs \rightarrow$ Photochemical oxidants $\rightarrow$ photochemical smog

2. Gasoline, lighter fluid, dry-cleaning fluids, oil-based paints, perfumes

Study the Figure: Figure 47.3

1. SO_2 (sulfur dioxide) and NO_x (Nitrogen oxides) in the presence of oxygen turn into H_2SO_4 (Sulfuric acid) $\rightarrow 2H^+ + SO_4^{2-}$

    SO_2 (Sulfur dioxide) and NO_x (Nitrogen oxides) in the presence of oxygen turn into HNO_3 (Nitric acid) $\rightarrow H^+ + NO_3^-$, which causes acid deposition.

2. Acid deposition has a number of environmental impacts:
    *   It lowers the pH of lakes and rivers which can impact aquatic organisms.
    *   It leads to deforestation which can lead to a loss of biodiversity.
    *   Metals bound in organic or inorganic compounds in soils and sediments are released into surface water which can impair the physiological functioning of aquatic organisms, exposure can lead to species loss.

Review Key Terms

1. b
2. a

Module 48: Pollution Control Measures

Practice the Math: Calculating Annual Sulfur Reductions

1. (384 ppm - 376 ppm) ÷ 376 ppm × 100% = 2%
2. Total percentage increase:
    (19.72 metric tons - 19.2 metric tons) ÷ 19.2 metric tons × 100% = 2.7% increase

    Annual percentage reduction:
    2.7% ÷ 15 years = 0.18 %/year

Module 50: Indoor Air Pollution

Review Key Terms

1. b
2. a

Chapter 15 Review Exercises

Check Your Understanding

1.

Compound	Effects
Sulfur dioxide	Respiratory irritant, damage plant tissue, harmful to aquatic life
Nitrogen dioxide	Respiratory irritant, forms photochemical smog, harmful to aquatic life, over fertilizes land and water
Carbon monoxide	Interferes with oxygen transport, headaches, can cause death
Particulate matter	Respiratory and cardiovascular disease, reduced lung function, contributes to haze and smog
Lead	Impairs central nervous system, effects learning and concentration
Ozone	Reduces lung function, degrades plant surfaces, damages rubber and plastic
VOCs	Precursor to ozone formation
Mercury	Impairs central nervous system, bioaccumulates up the food chain
Carbon dioxide	Affects climate, greenhouse gas

2. Photochemical smog is dominated by oxidants such as ozone. Sulfurous smog is dominated by sulfur dioxide and sulfate compounds.

3. Anthropogenic causes of air pollution include transportation, electricity generation, natural and human-made fires, and road dust.

4. A thermal inversion occurs when a warm layer traps emissions that then accumulate beneath the inversion layer. Pollution events created by a thermal inversion are particularly common in some cities, where high concentrations of vehicle exhaust and industrial emissions are easily trapped.

5. The effects of acid deposition include lowering the pH of lake water, decreased species diversity of aquatic organisms, mobilizing metals in soils and surface water, changes in trophic levels, damage to plants, and damage to human-built structures such as statues, monuments, and buildings.

6. Some locations, such as Mexico City, restrict the use of automobiles Some cities in Europe are experimenting with charging individual user fees to use roads at certain times of the day.

7. UV-C radiation breaks the bonds holding together the oxygen molecule, leaving two free oxygen atoms. The result is that in the presence of UV radiation, oxygen is converted to ozone. Ozone is broken down into O2 and free oxygen atoms. The free oxygen atoms and molecular oxygen may again react to produce ozone.

8. CFCs were used in refrigeration, air conditioners, propellants in aerosol cans, and in foam products like Styrofoam cups and foam insulation. CFCs were used in refrigeration, air conditioners, propellants in aerosol cans, and in foam products like Styrofoam cups and foam insulation. CFC's were banned because of the discovery of the link to the decrease in stratospheric ozone. It was discovered that when CFCs were introduced to the stratosphere, there was a rapid destruction of stratospheric ozone. In 1987, 24 nations signed the Montreal Protocol and committed to reduce CFC production by 50 percent by the year 2000.

9. Some indoor air pollutants are VOCs, asbestos, carbon monoxide, radon, tobacco smoke, paints and cleaning fluids.

Practice for Free-Response Questions

1. Primary pollutants are polluting compounds that come directly from a smokestack, exhaust pipe, or natural emission source. Examples of primary pollutants are VOCs, CO, CO_2, NO_2, NO, SO_2, most hydrocarbons, and most suspended particles.

2. Secondary pollutants are primary pollutants that have undergone transformation in the presence of sunlight, oxygen, water, or other compounds combine to form secondary pollutants. Examples of secondary pollutants are: SO_3, H_2SO_4, HNO_3, O_3, H_2O_2, most NO_3-, and SO_4^{2-}.

Chapter 16

Module 51: Only Humans Generate Waste

Study the Figure: Figure 51.5

1. 1.148 million metric tons ÷ 227 million metric tons = 65.2 %

2. Reasons that only a fraction of compostable MSW is actually composted include apathy, lack of understanding about how to compost, no place to compost, and accidentally throwing away compostable items

Review Key Terms

1. b   3. a
2. c

Module 52: The Three Rs and Compositing

Review Key Terms

1. b   3. a   5. f   7. d
2. e   4. g   6. f

Module 53: Landfills and Incineration

Study the Figure: Figure 53.2

1. 1. Solid waste is transported to landfill.
    2. Waste in compacted by a specialized machine.
    3. Leachate collection system removes water and contaminates and carries them to a waste water treatment plant.
    4. Landfill is capped and covered with soil and then planted with vegetation.
    5. Methane produced in closed cells is extracted and either burned off or collected for use as fuel.
2. Systems that are put into place in a sanitary landfill to stop leachate from contaminating groundwater include the use of sand, clay liners, gravel, leachate collection systems, groundwater monitoring wells.

Review Key Terms

1. f	4. e	7. i
2. d	5. h	8. b
3. g	6. a	9. b

Module 54: Hazardous Waste

Review Key Terms

1. c
2. a
3. b

Module 55: New Ways to Think About Solid Waste

Practice the Math

1. (a) 250 mm/year = 0.250 m/year
    0.250 m/year × 10,000 m^2 × 50%= 1250 m^3

    (b) 1250 m^3 × 80%= 1000 m^3

2. 100 mm/year = 0.1 m/year
    0.1m/year × 5000 m^2 = 500 m/year
    500 m/year × 40% = 500 m/year × 0.40 = 200 m^3
    200 m^3 × 70% = 200 m^3 × 0.70 = 140 m^3

Review Key Terms

1. a
2. b

Chapter 16 Review Exercises

Check Your Understanding

1. Organic material such as wood, yard trimmings, food scraps, and paper make up 64 percent of MSW, glass makes up 5 percent, metals percent, plastics 13 percent, rubber, leather, and textiles 8percent, and other 1 percent.

2. Parts of a landfill include leachate collection systems, a liner with sand and clay, and a method to monitor the groundwater for any leaks

3. Leachate that might contaminate underlying and adjacent waterways, risk to human health, and explosion hazards from the methane buildup. Sanitary landfills are constructed with a clay or plastic lining at the bottom. Clay is often used because it can impede water flow and retain positively charged ions, such as metals. A system of pipes is constructed below the landfill to collect leachate, which is sometimes recycled back into the landfill. Finally, a cover of soil and clay, called a cap, is installed when the landfill reaches capacity.

4. Problems with waste incineration include: ash that is left is more concentrated and more toxic, the cost of the practice, and air pollution.

5. RCRA (Resource Conservation and Recovery Act) was created to protect human health and the natural environment by reducing or eliminating the generation of hazardous waste. RCRA is known as "cradle to grave" tracking. CERCLA (Comprehensive Environmental Response, Compensation, and Liability Act) is also known as Superfund and authorizes the federal government to respond directly to the release or threatened release of substances that may pose a threat to human health or the environment.

6. Integrated Waste Management is an approach to waste disposal that employs several waste reduction, management, and disposal strategies in order to reduce the environmental impact of MSW. These consist mainly of reduce, reuse, and recycle. The phrase Integrated Waste Management, incorporates a practical approach to the subject of solid waste management, with each technique presented in the order of benefit to the environment, from the most desirable to the least.

Practice for Free-Response Questions

E-Waste is one component of MSW that is small by weight but very important and rapidly increasing. Consumer electronics that include televisions, computers, portable music players, and cell phones account are examples of e-waste. Heavy metal lead, as well as other toxic metals such as mercury and cadmium eventually leaches out of the bottom of the landfill into groundwater or surface water. In addition, much e-waste from the United States is exported to China where the materials are separated using fire and acids in open spaces with no protective clothing and no respiratory gear.

Chapter 17

Module 56: Human Diseases

Study the Figure: Figure 56.1

1. Cardiovascular disease and infectious diseases accounts for over 50 percent of all deaths in the world.

2. Many diseases have been eradicated worldwide by improvements in nutrition, wider availability of clean drinking water, proper sanitation, better healthcare, and education.

Study the Figure: Figure 56.6

1. $\dfrac{1.5 \text{ Million} - 2.5 \text{ million}}{2.5 \text{ million}} \times 100\% = -40\%$ (or a 40 percent decrease in tuberculosis deaths)

2. In the United States and other developed nations, antibiotics are readily available and healthcare professionals monitor compliance. These factors not only reduce the spread of tuberculosis, they also reduce the development of antibiotic resistant strains.

Review Key Terms

1. q	6. m	11. e	16. i
2. k	7. b	12. r	17. p
3. d	8. h	13. o	18. a
4. c	9. s	14. g	19. l
5. f	10. j	15. n	

Module 57: Toxicology and Chemical Risks

Practice the Math: Estimating LD50 Values

$$\frac{LD50}{100} = \text{assumed safe level for dogs}$$

$$\frac{2\text{mg/kg of body mass}}{100} = 0.02 \text{ mg/kg of body mass}$$

Review Key Terms

1. n	6. g	11. o	16. q
2. r	7. m	12. h	17. f
3. t	8. s	13. i	18. d
4. j	9. p	14. k	19. l
5. c	10. a	15. b	20. e

Module 58: Risk Analysis

Review Key Terms

1. c	3. e	5. b
2. a	4. d	

Chapter 17 Review Exercises

Check Your Understanding

1. Health risks of people living in developing nations include poverty, unsafe drinking water, poor sanitation, and malnutrition. Health risks of people living in developed nations include tobacco use, less active lifestyles, poor nutrition, and overeating that leads to high blood pressure and obesity.

2.

Disease	How Disease is Spread
Plague	Fleas on rats
Malaria	Mosquitoes
Tuberculosis	Person to person
HIV/AIDS	Person to person
Ebola hemorrhagic fever	Person to person
Mad cow disease	Eating meat from contaminated cattle
Bird flu	Birds to people
West Nile virus	Mosquitoes

3. Some different pathways include domesticated animals, other humans, water, air, food, and wild animals.

4. See Table 57.1.

5. LD50 refers to the amount of a specific toxin required to kill 50 percent of individuals exposed to it. ED50 describes how much of a toxin is required for 50 percent of individuals exposed to a toxin to experience harmful effects, but not death.

6. DDT is a persistent chemical that increases in concentration as it moves up the food chain. As DDT moved up the food chain it was stored in the fat of the animals that ate it and eventually made its way up to the predatory birds. These birds had eggs that were thin-shelled and often broke when the parent bird incubated the eggs.

7. When conducting risk analysis one must identify the hazard, characterize the toxicity (does/response), and determine the extent of exposure. This is followed by determining the acceptable level of risk, and finally deciding if a policy should be enacted with input from private citizens, industry, and interest groups.

Practice for Free-Response Questions

Bioaccumulation occurs when an organism increases the concentration of a chemical in its body over time. Biomagnification is the increase in the chemical concentration in animal tissues as the chemical moves up the food chain. An example of bioaccumulation was when DDT accumulated in the fatty tissues of animals. Biomagnification ocurred when DDT worked its way up the food chain, increasing chemical concentrations at each level.

Unit 7 Multiple-Choice Review Exam

1. D	8. D	15. A	22. C
2. A	9. D	16. E	23. B
3. E	10. C	17. C	24. E
4. B	11. A	18. D	25. D
5. C	12. E	19. E	26. E
6. A	13. B	20. B	27. B
7. B	14. A	21. A	28. E

Chapter 18

Module 59: The Sixth Mass Extinction

Study the Figure: Figure 59.4

1. Birds have had the least percentage of global declines since the year 1500.

Review Key Terms

1. d	3. f	5. c	7. e
2. a	4. g	6. b	

Module 60: Causes of Declining Biodiversity

Study the Figure: Figure 60.2

1. The portion of the world that has seen the greatest increase in forest cover is the United States.

Review Key Terms

1. e	3. a	5. c
2. b	4. f	6. d

Module 61: The Conservation of Biodiversity

Review Key Terms

1. a	3. b	5. f
2. c	4. e	6. d

Chapter 18 Review Exercises

Check Your Understanding

1. The five categories used by the International Union for Conservation of Nature are data deficient, extinct, threatened, near-threatened, and least concern.

2. Invasive species have no natural enemies to control their population. They pose a serious threat to biodiversity by acting as predators, pathogens, or superior competitors to native species. Examples of invasive species include rats, exotic fungi, kudzu vine, zebra mussel, and silver carp.

3. Lacey Act: The earliest law to control trade of wildlife.
   CITES: An international treaty that controls the international trade of threatened plants and animals.
   Marine Mammal Protection Act: An act that prohibits the killing of all marine mammals in the United States and prohibits the import or export of any marine mammal body part.
   Endangered Species Act: An act that prohibits the harming of any threatened or endangered species including trade of the species. It also gave the government the right to purchase habitats that are critical to the survival of these species.
   Convention on Biological Diversity: An international treaty to help protect biodiversity.

4. Biosphere reserves consist of core areas that have minimal human impact and outer zones that have increasing levels of human impacts. The reasons behind biodiversity decline include habitat loss, exotic species, overharvesting, pollution, climate change.

5. Possible answers include conservation legislation, protecting ecosystems, and prevention of habitat loss, invasion of exotic species, overharvesting, pollution, and climate change.

Practice for Free-Response Questions

The Convention on Biological Diversity is an international treaty to help protect biodiversity. Trends include:
- On average species at risk of extinction have moved closer to extinction.
- One-quarter of all plant species are still threatened with extinction.
- Natural habitats are becoming smaller and more fragmented.
- The genetic diversity of crops and livestock is still declining.
- There is a widespread loss of ecosystem function.
- The causes of biodiversity loss have either stayed the same or increased in intensity.
- The ecological footprint of humans has increased.

Chapter 19

Module 62: Global Climate Change and the Greenhouse Effect

Study the Figure: Figure 62.2

1. 1. Incoming solar radiation consists primarily of UV and visible light.

    2. About 1/3 of the solar radiation is reflected—from the atmosphere, clouds, and the surface of the planet—back into space.

    3. The remaining solar radiation is absorbed by clouds and the surface of the planet both become warmer and then emit infrared radiation.

    4. Much of the emitted infrared radiation from earth is absorbed by greenhouse gases in the atmosphere. The remainder is emitted into space.

    5. As the greenhouse gases absorb infrared radiation, they warm and emit infrared radiation, with much of it going back towards earth. The greater the concentration of greenhouse gases, the more infrared radiation is absorbed and emitted back toward Earth.

2. Some answers include: ice caps melt, glaciers melt, permafrost melts, sea level rises, estuaries can flood, thermal expansion of ocean water, stronger storms, droughts, floods, impacts on organisms.

Study the Figure: Figure 62.6

1. Energy combustion and production are responsible for the largest contribution of methane.

    2. Synthetic fertilizers, manures, and crops that naturally fix atmospheric nitrogen can create an excess of nitrates in the soil that are converted to nitrous oxide by the process of denitrification.

    3. The two anthropogenic sources are combustion to generate electricity (38 percent) and combustion for transportation (31 percent.)

Review Key Terms

1. c   3. e   5. d
2. a   4. b

Module 63: The Evidence for Global Warming

Study the Figure: Figure 63.1

1. $\dfrac{380\text{-}360}{360} \times 100\% = 5.56\%$

2. Reasons for the steady rise in atmospheric CO_2 include automobiles, transportation, combustion for homes and businesses, industrial processes, burning fossil fuels.

Practice the Math: Projecting Future Increases in CO_2

1. 1. 1.4 ppm × 40 years = 56 ppm
390 + 56 = 466 ppm in 2050

3. 1.9 × 40 years = 76 ppm
390 + 56 = 446 ppm

4. Years: 2150 − 2010 = 140 years
140 years (1.4 ppm CO_2 /year) = 196 ppm extra CO_2

Predicted concentration in 2150 = concentration 2010 + calculated extra
= 390 ppm + 196ppm
= 586 ppm CO_2

Module 64: The Consequences of Global Climate Change

Review Key Terms

1. b
2. a

Chapter 19 Review Exercises

Check Your Understanding

1. When the high-energy radiation from the Sun strikes the atmosphere, about one-third is reflected from the atmosphere, clouds, and the surface of Earth. Much of the high-energy radiation is absorbed by the ozone layer where it is converted to low-energy infrared radiation. Some of the ultraviolet radiation and much of the visible light strikes the land and water of Earth where it is also converted into low-energy infrared radiation. The infrared radiation radiates back toward the atmosphere where it is absorbed by greenhouse gases that radiate much of it back toward the surface of Earth. Collectively, these processes cause warming of the planet.

2. See Table 62.1, page 669.

3. In the spring, deciduous trees, grasslands, and farmlands in the Northern Hemisphere turn green, so they increase their carbon dioxide intake for photosynthesis. In the winter, they lose their leaves and do not take in as much carbon dioxide.

4. When scientists melt pieces of ice they extract from drilling tubes of ice, they analyze the air bubbles that are released by measuring the concentration of greenhouse gases. They can then determine temperature changes over long periods of time.

5. Two possible explanations for warming temperatures on Earth are an increase in solar radiation and the possibility that warming is caused by increased carbon dioxide levels in addition to normal fluctuations.

6. Carbon sequestration is an approach to taking carbon dioxide out of the atmosphere. This might include storing carbon in agricultural soils or retiring agricultural land and allowing it to become pasture or forest. Researchers are also trying to find ways of capturing carbon dioxide from coal-burning power stations.

Practice for Free-Response Questions

A volcanic eruption emits a large quantity of ash into the atmosphere. The ash reflects incoming solar radiation back out into space, which has a cooling effect on Earth.

Chapter 20

Module 65: Sustainability and Economics

Study the Figure: Figure 65.1

1. At the equilibrium point, supply equals demand. At the corresponding price, there is neither a shortage nor a surplus of the good.

Study the Figure: Figure 65.4

1. GPI initially increases rapidly when GDP first increases but then reaches an apex and begins to decline.

2. Some things that could be put into place to improve the environment are scrubbers on power plants, pollution laws, catalytic converters on vehicles, education, and taxing polluters.

Review Key Terms

1. i	4. a	7. e	10. c
2. f	5. d	8. j	11. c
3. h	6. l	9. b	12. g

Module 66: Regulations and Equity

Review Key Terms

1. d	6. b	11. n	16. q
2. g	7. e	12. o	17. m
3. l	8. r	13. f	18. k
4. p	9. c	14. j	19. h
5. a	10. i	15. s	

Chapter 20 Review Exercises

Check Your Understanding

1. Supply is the number of units of an item a manufacturer will provide. Demand is the amount of units of an item consumers want.

2. The goals are to eradicate extreme poverty and hunger, achieve universal primary education, promote gender equality and empower women, reduce child mortality, improve maternal health, combat HIV/AIDS, malaria, and other diseases, ensure environmental sustainability, and develop a global partnership for development.

3. Dr. Wangari Maathai founded the Green Belt Movement, a Kenyan and international environmental organization that empowers women by paying them to plant trees, some of which can be harvested for firewood in a few years.

Practice for Free-Response Questions

As GDP grows, population growth slows. This in turn, should lead to a reduction in anthropogenic environmental degradation. In addition, wealthier, developed nations are able to use economic resources to protect and improve the environment. For example, more developed nations are more likely to use pollution control devices such as catalytic converters. Wealthier nations have a more stable political system that allows for legislation of environmental protection and the enforcement of such legislation. Wealthier nations can also invest in renewable resource technology such as wind and solar.

Unit 8 Multiple-Choice Review Exam

1. B	7. C	13. B	19. D
2. E	8. A	14. D	20. E
3. A	9. C	15. A	21. E
4. D	10. E	16. E	
5. C	11. B	17. B	
6. B	12. C	18. C	

Full-Length Practice Exam 1

SECTION I: Multiple-Choice

1. C	26. C	51. A	76. B
2. A	27. A	52. A	77. C
3. C	28. B	53. E	78. A
4. A	29. A	54. A	79. C
5. D	30. C	55. E	80. C
6. D	31. B	56. D	81. C
7. B	32. B	57. C	82. E
8. C	33. D	58. A	83. B
9. D	34. D	59. B	84. E
10. B	35. C	60. E	85. B
11. D	36. A	61. D	86. A
12. D	37. B	62. B	87. E
13. C	38. E	63. A	88. C
14. D	39. A	64. B	89. A
15. A	40. A	65. D	90. B
16. A	41. D	66. C	91. E
17. D	42. E	67. B	92. D
18. A	43. C	68. B	93. A
19. C	44. A	69. D	94. A
20. D	45. B	70. E	95. C
21. D	46. B	71. A	96. B
22. E	47. A	72. C	97. D
23. A	48. C	73. B	98. E
24. C	49. B	74. A	99. B
25. A	50. C	75. C	100. B

SECTION II: Answers to Free-Response Questions

1. (A) Farmers may choose to grow these crops because they reduce the need for pesticides and increase crop yields. These two factors increase profits for the farmer. A farmer may choose to grow GMO crops in areas where soil quality is poor if the engineered plant is better adapted to these conditions. Lastly, a farmer may choose to grow nutritionally dense foods such as golden rice to benefit human health.

    (B) Scientists isolate a specific gene from one organism and transfer it into the genetic material of another, often very different, organism.

    (C)
    - The chance of an allergic reaction
    - The effects on biodiversity
    - The lack of regulation

    (D)
    - Intercropping: Two or more crop species are planted in the same field at the same time to promote a synergistic interaction between them.
    - Crop rotation: The crop species in a field are rotated from season to season.
    - Agroforestry: Vegetation of different heights, including trees, act as windbreaks and catch soil that might otherwise be blown away, and greatly reduces erosion.
    - Contour plowing: Plowing and harvesting are done parallel to the topographic contours of the land, which helps prevent erosion by water while still allowing for the practical advantages of plowing.
    - No-till agriculture: Crop residues in the field are left between seasons so that the roots hold the soil in place, reducing both wind and water erosion

2. (A) First, determine how many gallons in 1 cubic foot.

    748 gallons in 100 cubic feet

    $748 \div 100 = 7.48$ gallons

    Convert 8,000 gallons to cubic feet.

    $8,000 \div 7.48 = 1069.52$ cubic feet

    Determine cost of water.

    100 cubic feet = $3.45

    $(1069.52 \text{ cubic feet} \div 100) \times \$3.45 = \$36.90$

(B) 42 gallons per shower ÷ 2 = 21 gallons with new shower head
21 gallons per shower × 4 people = 84 gallons per day

84 gallons per day × 7 days = 588 gallons per week

(C)
- Low flow toilet,
- Larger loads of dishes or laundry,
- Reuse gray water
- Turn off the water when you brush your teeth
- Fix leaks
- Don't let water run as you do dishes

(D)
- Don't water as often
- Use a programmable sprinkler system
- Water in the evening or at night
- Use drip irrigation
- Landscape with plants that do not need much water

3. (A) CFCs and other chemicals rise on air currents into the stratosphere and break down the ozone.

(B) CFCs released in the troposphere travel to the stratosphere. Once in the stratosphere CFCs are exposed to UV radiation which releases a chlorine atom. The chlorine atom removes an oxygen atom from ozone creating ClO and diatomic oxygen. ClO is relatively unstable and the oxygen will dissociate from ClO and combine with free oxygen atoms in the stratosphere creating diatomic oxygen. The free chlorine can then attack another ozone molecule and the process will be repeated.

$$CFCl_4 + UV \rightarrow CFCl_3 + Cl$$

$$Cl + O_3 \rightarrow ClO + O_2$$

$$ClO + O \rightarrow O2 + Cl$$

(C)
- Harmful to plant cells
- Can reduce photosynthetic activity
- Lowers ecosystem productivity
- Increases risk of skin cancer
- Causes cataracts and other eye problems
- Suppresses immune system

(D) Montreal Protocol

(E)
- Asthma
- COPD
- Emphysema
- Breathing difficulties
- Cancer

We are burning fossil fuels and burning biomass.

4.

(A)
- **Phase 1:** Population is stable. Births and deaths are both high and life expectancy is relatively short. Infant mortality rate is high and there is pore sanitation. relatively short. Infant mortality rate is high and there is pore sanitation.
- **Phase 2:** Death rates decline and birth rates remain high so the population is growing rapidly. There is better sanitation, clean drinking water, increased access to food and healthcare so the infant mortality rate decreases.
- **Phase 3:** The birth rate falls and the population becomes stable. This country is developed and is relatively affluent. Education is high and the availability of birth control increases.
- **Phase 4:** The population begins to decline. There is a high level of affluence and economic development. There are fewer young people and higher numbers of elderly. There might be a shortage of workers.

(B) $\dfrac{(100 + 3\,0) - (80 - 20)}{10,000} = .7\%$

(C)
- The family needs help with work on the farm or in the home.
- There is a high infant and child mortality rates.
- Women do not have access to reliable birth control.

(D) $70 \div 0.7 = 100$ year

Full-Length Practice Exam 2

SECTION I: Multiple-Choice

1. E	27. B	53. E	79. E
2. C	28. E	54. A	80. A
3. E	29. D	55. C	81. D
4. A	30. E	56. B	82. B
5. C	31. B	57. D	83. D
6. B	32. A	58. D	84. D
7. E	33. B	59. E	85. E
8. E	34. B	60. C	86. E
9. A	35. E	61. A	87. C
10. E	36. C	62. D	88. D
11. B	37. C	63. C	89. E
12. A	38. A	64. A	90. E
13. A	39. C	65. B	91. D
14. C	40. B	66. E	92. A
15. C	41. E	67. A	93. A
16. B	42. A	68. C	94. A
17. C	43. A	69. A	95. A
18. A	44. D	70. B	96. E
19. A	45. A	71. E	97. D
20. E	46. B	72. B	98. B
21. B	47. D	73. E	99. B
22. D	48. A	74. A	100. E
23. E	49. D	75. E	
24. D	50. A	76. B	
25. C	51. A	77. E	
26. A	52. D	78. E	

SECTION II: Answers to Free-Response Questions

1. (A) Nitrates and phosphates would create an algae bloom which could create a dead zone or less dissolved oxygen; high fecal bacteria could change the types of organisms that live in the local river; viruses could lead to an increased presence of leeches and black fly larva.

   (B) **Primary treatment:** Solid waste is removed from the water (physical treatment).
   Secondary treatment: Bacteria is used to break down the organic matter in the water (biological treatment).

(C) Animal feedlots are places where large numbers of animals live in one place. They contain animal waste that is often full of hormones and antibiotics. The livestock operation could use manure lagoons, which are large, human-made ponds that are lined with rubber to prevent the manure from leaking out.

(D) Point source comes from a particular place such as a factory. Nonpoint comes from large areas such as farming regions, lawns, and storm runoff.

2. (A) 20,000 kWh per month × $.10 per kWh = $2,000

(B) 2000 kWh per month × $.10 = $200.00 per month
$15,000 ÷ $200 = 75 months

75 months ÷ 12 months per year = 6.25 years (or 6 years and 3 months)

(C) $15,000 of solar panels produce 2000 kWh, so $30,000 of panels will produce 2000 × 2 = 4000 kWh

(D)
- Turn off lights when not in use
- Insulate
- Use less air conditioning or heating
- Turn air off after school hours
- Keep doors closed, use motion sensor lights
- Use natural lighting (sunroof, etc.)

(E)
- Put curtains or blinds on windows
- Plant trees outside windows that face the sun
- Install a living roof
- Install awnings over windows
- Install a thermal mass near the windows to heat the space during the winter months

3.
(A)
- Carbon dioxide: burning of fossil fuels, deforestation, forest fires
- Water vapor: natural but increasing due to warmer temperatures
- Methane: cattle farming, sewage treatment plants, landfills, termites
- CFCs: air conditioners, aerosols, coolants, industrial processes
- NO_x: vehicle exhaust and fossil fuel combustion

(B) The CFCs released in the troposphere travel to the stratosphere. Once in the stratosphere CFCs are exposed to UV radiation which releases a chlorine atom. The chlorine atom removes an oxygen atom from ozone creating ClO and diatomic oxygen. ClO is relatively unstable and the oxygen will dissociate from ClO and combine with free oxygen atoms in the stratosphere creating diatomic oxygen. The free chlorine can then attack another ozone molecule and the process will be repeated.

$$CFCl_4 + UV \rightarrow CFCl_3 + Cl$$

$$Cl + O_3 \rightarrow ClO + O_2$$

$$ClO + O \rightarrow O_2 + Cl$$

(C)
- Glaciers melting
- Ice at the poles melting
- Sea levels rising
- Loss of biodiversity

(D)
   i. The troposphere is the layer of the atmosphere where these gases cause climate change.
   ii.
- Driving less
- Carpooling
- Riding a bike
- Eating less meat
- Leaving large tracts of forest untouched,
- Burning off methane from a sewage treatment plant or landfill

4.

(A) Photochemical smog is an air pollutant that formed as a result of sunlight acting on compounds such as nitrogen oxides and sulfur dioxide (also called Los-Angeles type smog or brown smog).

Sulfurous smog (also called gray smog or London-type smog) is dominated by sulfur dioxide and sulfate compounds and produces a brown cloud. This comes from the combustion of fossil fuel and burning biomass.

(B) Smog harms plant tissues, exacerbates human respiratory problems such as asthma. It is harmful to construction materials, causes poor visibility in popular vacation destinations, thereby reducing tourism revenue.

(C) Sulfur dioxide and nitrogen dioxide are released into the air from industrial plants. In the atmosphere they are converted to sulfuric acid and nitric acid and can fall to the earth in the form of wet-acid in rain and snow or dry-acid in gases and particles.

(D) Acid deposition can decrease the pH of lake water, causing the loss of delicate species. It can mobilize metals that are bound in soils and sediments and release them into surface water. In addition, acid deposition affects the food sources of aquatic organisms, is harmful to some species of trees, and erodes human-built structures such as statues, monuments, and buildings.